Transylvanian Dinosaurs

Transylvanian Dinosaurs

DAVID B. WEISHAMPEL and
CORALIA-MARIA JIANU

The Johns Hopkins University Press
Baltimore

© 2011 The Johns Hopkins University Press
All rights reserved. Published 2011
Printed in the United States of America on acid-free paper
9 8 7 6 5 4 3 2 1

The Johns Hopkins University Press
2715 North Charles Street
Baltimore, Maryland 21218-4363
www.press.jhu.edu

Library of Congress Cataloging-in-Publication Data
Weishampel, David B., 1952–
Transylvanian dinosaurs / by David B. Weishampel and Coralia-Maria Jianu.
    p. cm.
  Includes bibliographical references and index.
  ISBN-13: 978-1-4214-0027-3 (hardcover : alk. paper)
  ISBN-10: 1-4214-0027-8 (hardcover : alk. paper)
   1. Dinosaurs—Romania—Transylvania. 2. Dinosaurs—Evolution.
I. Jianu, Coralia-Maria. II. Title.
QE861.9.R62T739 2011
567.909498'4—dc22 2010047481

A catalog record for this book is available from the British Library.

All artwork by D. B. Weishampel, unless otherwise indicated.

Special discounts are available for bulk purchases of this book. For more information, please contact Special Sales at 410-516-6936 or specialsales@press.jhu.edu.

The Johns Hopkins University Press uses environmentally friendly book materials, including recycled text paper that is composed of at least 30 percent post-consumer waste, whenever possible.

The gift of life, it's a twist of fate
It's a roll of the die
It's a free lunch, a free ride
But Nature's got rules and Nature's got laws
And if you cross her, look out!

Laurie Anderson, *Strange Angels* (1989)

What follows is dedicated to Jack Horner.

CONTENTS

Acknowledgments   ix

CHAPTER 1. Bringing It All Back Home   1
CHAPTER 2. Dinosauria of Transylvania   15
CHAPTER 3. Pterosaurs, Crocs, and Mammals, Oh My   59
CHAPTER 4. Living on the Edge   84
CHAPTER 5. Little Giants and Big Dwarfs   113
CHAPTER 6. Living Fossils and Their Ghosts: Being a Short Interlude on Coelacanths and Transylvanian Ornithopods   147
CHAPTER 7. Transylvania, the Land of Contingency   163
CHAPTER 8. Alice and the End   194

Notes   203
Glossary   221
References   241
Index   291

Color plates follow page 112.

ACKNOWLEDGMENTS

Books like this do not spring fully formed from the forehead of Jove. Rather, they are most often inspired by the work of others, coming together from fragments of old and new ideas that careen from right brain to left, scrambling and unscrambling through international mail systems and e-mails. Parts die and parts survive as the goals of the book become more tangible.

Little could he have known it, but Franz Baron Nopcsa and his quest to understand the dinosaurs emanating from his backyard have led this American and this Romanian to contemplate how the Transylvanian dinosaurs fit into the Mesozoic scheme of things. Perhaps he'd be pleased by our efforts to ask big questions that demand eclectic and multidisciplinary answers. The Baron had a lot to say in his relatively short life, and we've found that it pays to listen to him.

Our studies have taken us many places as we followed in Nopcsa's footsteps, investigated new dinosaur material in Europe and elsewhere, and stumbled back and forth through the dust-coated central European evolutionary biology of the early twentieth century and the twists and turns of contemporary systematics, developmental biology, and evolutionary theory. In doing so, we have encountered many friends, fellow sojourners, and mentors, all of whom we would like to thank here.

We begin by thanking the members of our field crews—especially Cristi and Puiu from Sânpetru, Bogdan Scarlat (Bob), Ovidiu Hebler (Ova), Dan Șuiagă (Șușu), Cristina Circo, and Sebastian Domnariu; the Italians Fabio Dalla Vecchia, Davide Rigo, and Cezare Brizio; and, of course, the international team, mostly Dutch: Anne Schulp, Remmert Schouten, Matthew Deeks, Moritz von Graevenitz, Jac and Emile Philippens, and Roos de Klerk—for their hard work, most excellent partying, and camaraderie. Mulțumim, li ringraziamo, dank u wel! And a special köszönöm to Zoltán Csiki, master of the Bucharest bones, for his exquisite knowledge, his friendship, and his ever-present "yes, but . . ."

Mulţumim also to other Romanian colleagues: to Dan Grigorescu from Universitatea din Bucureşti who originally invited the senior author to Romania and shared a few field seasons with him; to Ioan Groza (now deceased) for establishing the dinosaur collections at Muzeul Civilizaţiei Dacice şi Romane Deva; to Doinel Vulc, the most able paleontologist in Sânpetru; to Nicolae Mészaros (now deceased) and Vlad Codrea from Universitatea Babeş-Bolyai in Cluj-Napoca for their own work on the Transylvanian assemblages and also for their help; and to Costin Rădulescu and Petre Samson (both now deceased) from Institutul Speologic "Emil Racoviţă" in Bucharest for making it possible for us to study their spectacular multituberculate material and for their encouragement.

For the possibility of examining the Transylvanian dinosaurs and the Nopcsa archives in their respective institutions, as well as for the pleasure of enjoying summertime in Budapest with them, we thank László Kordos and Jozef Hála from the Magyar Állami Földtani Intézet, and István Fözy and Horváth Csaba from the Magyar Természettudományi Muzeum. Nagyon szépen köszönöm!

Nonetheless, as Nopcsa would have wanted it, the best of all collections of the Late Cretaceous dinosaurs from Transylvania still resides in London, where they were originally presented to the (then) British Museum (Natural History) in 1906 and 1923. At his death, Nopcsa's paleontological archives were bequeathed to this same institution. Angela C. Milner and Sandra Chapman now mind the Nopcsa fossil and archival collections, and we thank them for the opportunity to feast upon the fossils and memorabilia, including Nopcsa's "brain."

Thanks also go to our French colleagues Jean Le Loeuff, Eric Buffetaut, Marie Pincemaille, Yves Laurent, François Sirugue, and Patrick Mechan, mostly from Espéraza, Montpellier, and Aix-en-Provence, for the best cooking and wine that the great land of southern France offers and for sharing with us the Late Cretaceous riches they have gleaned from the rocks at Bellevue, Corbières, and Le Mas d'Azil. Jean Le Loeuff also provided the inspiration for the title of chapter 5. Merci, tout le monde!

We especially thank Angela D. Buscalioni and Francisco Ortega from the Universidad Autónoma de Madrid for their wisdom on things crocodilian, for their friendship, and for the ride to Dalí's hometown—muchas gracias y viva Figueres! We also thank Xabier Pereda-Suberbiola from the Universidad del País Vasco / Euskal Herrikom Unibertsitatea, Bilbao, and the Université Paris VI for his work on *Struthiosaurus* and other European ankylosaurs and for his interest in the life and times of Franz Nopcsa. Eskerrik asko, Xabier.

Vienna, Nopcsa's abode for most of his life, including his university years, is the home of important dinosaur material from the Gosau Beds in eastern Austria. Gernot Rabeder, Norbert Vávra, Doris Nagel, and Karl Rauscher from Universität Wien provided us with the opportunity to study this material and also forged our connections with the various archives in Vienna, which contain the most extensive records of Nopcsa's life. Vielen dank für alle Ihre Hilfe.

Last, but certainly not least, we are grateful to Jiří Kvaček at the Národní Muzeum in Prague, Kafka's city of dark contingency, for enabling us to examine the pterosaurs and other fossil material in his care. Děkuju mockrát.

Looking beyond the field and museum collections that immediately relate to Transylvanian dinosaurs and the Baron, there are many people who have helped us to explore the nature of history and to see the tapestry we were constructing from new perspectives. Thanks go to Anne Schulp (again), John Jagt, and Douwe Th. de Graaf (the "Maastricht Guys"), and Marcin Machalski from the Instytut Paleobiologii in Warsaw, for somehow putting up with our demanding, raucous, and oftentimes unpredictable behavior, all for the honorable sake of creating the traveling museum exhibition called "Dinosaurs, Ammonites, and Asteroids: Life and Death in the Maastrichtian." We also thank Eric Mulder from the Museum Natura Docet in Denekamp, the Netherlands, for keeping the light shining on the truly Maastrichtian dinosaurs, those from Maastricht.

To Dave Norman from the Sedgwick Museum, Cambridge University, thanks for your dinosaur knowledge. To Jack Horner from the Museum of the Rockies in Bozeman, Montana, thanks for your encouragement on the Transylvanian and dinosaurian fronts. To Oliver Kerscher for your ever-patient journey through our translation from German to English of Tasnádi-Kubacska's biography of Nopcsa, and to Robert Elsie, fellow traveler—but from the Albanian side—regarding Nopcsa's remarkable life, many thanks to you both.

We are especially grateful to the Interlibrary Loan staff at the Johns Hopkins University's Welch Library, especially Ellen Priebe and Tina Avarelli, for handling a bewildering array of requests—in English, Romanian, German, French, Hungarian, and Serbo-Croatian—on subjects ranging from earthquakes in Italy to Albanian ethnography to dinosaur-avian relationships. Arriving at incredibly regular intervals, these materials have proved to be indispensable to this project.

For financial assistance, we thank the generosity of the U.S. National Research Council, the U.S. National Science Foundation, the National Geographic Society, the Dinosaur Society, the Jurassic Foundation, the Paleontological Society International Research Program (Sepkoski Grants), and Muzeul Civilizației Dacice și Romane Deva (MCDRD). Together, they made possible ten field seasons in the Hațeg Basin, museum research on additional collections of the dinosaurs from Transylvania and elsewhere in Europe, and research at the many Nopcsa archives. We also thank Adriana Pescaru, retired general director, and Silvia Burnaz, head of Secția de Științele Naturii, MCDRD, for providing the institutional basis for the successes of the various field seasons conducted by the Joint MCDRD–Johns Hopkins University (JHU) Dinosaur Expeditions.

The National Geographic Society also made it possible for us to join Attila Ősi and his important work on the Late Cretaceous of Iharkút, Hungary. The remains of a host of wonderfully preserved dinosaurs, pterosaurs, crocodilians, and squamates are now coming out of the walls of an open-pit former bauxite mine, and, should this locality continue to yield as abundantly as it does at present, it will become one of the most significant in Europe.

Each of us, of course, has individuals to whom special thanks are due.

Dave Weishampel: First and foremost, I thank Cora Jianu—for your devil's advocate role as coauthor, for your considerable insight into the geologic, ecological, and evolutionary dynamics of the Transylvanian region and its fauna, and for inviting me to conduct this Transylvanian research with you. Mulțumesc foarte mult, prietena mea. Thanks also go to Wolf-Ernst Reif (deceased) from the University of Tübingen for his tutelage in constructional and theoretical morphology, as well as in the inherent tension between natural laws and contingency in creating history. To Ronald E. Heinrich, even though you may not know it, for hinting that this should not be just another dinosaur book. To friends at

JHU—Ken Rose, Mason Meers, Mary Silcox, Jason Mussell, François Therrien, Amy Chew, Tonya Penkrot, Shawn Zack, Matt O'Neil, Jason Organ, Ben Auerbach, Frank Varriale, and Mike Habib—many thanks for your efforts to answer all those bizarre or picky questions I threw at you over the past handful of years. Finally, special thanks go to Nachio Minoura, who made possible my Visiting Professorship at the University of Hokkaido Museum in Sapporo, Japan. I spent the winter months of 1999–2000 there, sliding on the ice to and from his office, enjoying wayward talks over coffee, and sampling the delights of hometown sushi, all the while exploring the worlds of ancient cultures, the delicacy of the natural world, and historical contingency. Dōmo arigatō, Nachio.

Cora Jianu: This is the place for special thanks to my family and friends, especially to my parents, my late grandmother, and my son, Ion—for putting up with all my time away conducting fieldwork and on other work-related travels, and for your understanding, support, and encouragement when I needed it. Thanks also to my friend Bert Boekschoten from Vrije Universiteit at Amsterdam for believing in me. To Professor Vlad Codrea at Universitatea Babeş Bolyai in Cluj-Napoca for early encouragement in my career as a paleontologist. To Professor Anton Fistani from the Universiteti i Shkodrës "Luigj Gurakuqi," Shkoder, Albania, for being such an accommodating host to me and my friends on our unannounced visit to Shkoder, following in the steps of Franz Nopcsa. It took some investigation to find the street named after the Baron, but we were very pleased when we did! Finally, to Dave Weishampel for teaching me about art, music, evolution, contingency, and cladistics. Dave, thank you for inviting me to write this book with you, and for not giving up on this project; it's been such an enlightening experience! With your superb illustrations, the book is very much enriched and brought to life.

Ron Heinrich, François Therrien, Zoli Csiki, and Mike Habib read early drafts of this book, and we are especially grateful for their insight, criticism, kindness, and forbearance. Their advice focused our prose, suggested directions, and provided many new contexts for our thoughts. Thanks.

Laurie Anderson kindly provided permission to reproduce some of the lyrics to "Monkey's Paw." Apologies go to Salvador Dalí for our "plagiarizing" a portion of his work, *The Temptation of St. Anthony*, on the dedication page.

Finally, we extend special thanks to the Johns Hopkins University Press, particularly to Vince Burke, for taking on this book and allowing us to expand the project in ways we couldn't have anticipated from the start. Ideas are the important thing, and Vince gave us the opportunity to weave the dinosaurs of Transylvania through the Late Cretaceous contingencies of the Eastern European world, touching upon art, film, and pinball as we proceeded. The challenge was to make our story clear, insightful, and fun, and for that we thank him for all his help.

Transylvanian Dinosaurs

CHAPTER 1

Bringing It All Back Home

Morning comes early when you're in the field. No matter where you are—Mongolia, Montana, or Madagascar—it's bound to be impossibly hot later in the day. So it's better to be looking for fossils when it's reasonably comfortable to be walking around or digging in the rocks to extract dinosaur bones and teeth.

It's 6:30 a.m. and we're camped on a small piece of real estate, a wooded island of sorts, midstream in the Sibişel River, which rushes by us from its headwaters in the Retezat Mountains some 25 kilometers away to the south, eventually to reach the Danube and then the Black Sea. We are in the northern foothills of the Southern Carpathian Mountains of Transylvania in western Romania.

It's clear and crisp this morning and has been dry for about a week now, ever since we began our fieldwork near the village of Sânpetru. The night's dew still clings to the outside of our domed tents, which make an assemblage of what looks like blue turtles (Plate I, top). The camp also has a few more-conventional walled tents for storage and a kitchen. From the cook tent, the smell of freshly brewed camp coffee floats across our small opening in the woods and others must smell it, too, since they are also groggily emerging from their tents.

In addition to the strong coffee, breakfast consists of bread and cheeses —pâine şi brânză—and some juice or fruits. Sometimes there is slănină, a kind of thick bacon nearly devoid of meat. Not much good for those of us who are vegetarians, but many enjoy its taste and the way it sizzles as we watch the sun come up.

After breakfast, it's time to pack up the bags of supplies we'll need in the field, making sure to bring water and checking that there is fuel for the jackhammer. After everything is accounted for, we head out of our camp. Single file, we cross the makeshift bridge of timber and ancient wooden slats to get to the west side of the Sibişel River and the dirt road that connects Sânpetru to the village of Ohaba Sibişel farther to the south, with exposures of mixed orange and gray-green rocks outcropping on our side of the river and along the bluffs on the other side. Women from the village pound out their laundry on the rocks along the banks of the Sibişel, while their husbands and older children tend their herds of sheep and cows in the hills above the river.

Prospecting and quarrying are the order of the day, so our crew (Plate I, bottom)—variable in number but often consisting of a dozen or so volunteers from nearby towns and cities, high schoolers as well as friends and colleagues—is split in two. Prospecting, the foundation of all paleontological discoveries, is the only way we can know where there are fossils—by walking and looking. On the whole, there are lots of bones to be found in the rock outcroppings in the Hațeg Basin in southern Hunedoara County, especially along a 2.5 km stretch of the Sibişel Valley directly south of Sânpetru. These rocks constitute the Sânpetru Formation, a cyclic stack of sandstones and mudstones that were originally deposited here as sediment by braided streams and floodplains that occupied the lower part of an alluvial fan about 70 million years ago, during the Late Cretaceous. A good time period and environment in Earth's history to encounter our ancient objectives: dinosaurs.

Locating the remains of dinosaurs and other extinct organisms, however, is generally a matter of serendipity. The problem in this part of the world is twofold: most of the fossils occur in isolation, with the separate bits removed from each other, and most of the rocks available for collecting are covered with vegetation. As a result, you need to have good eyes to see these bones and teeth emerging from their sedimentary tomb, as well as a strong back and legs to clamber up rocky slopes and range widely from exposure to exposure. Fortunately, we have both; prospecting thus far has yielded several isolated vertebrae belonging to two different kinds of dinosaurs (the ornithopod *Zalmoxes* and a titanosaurid sauropod), a cervical rib whose owner we haven't yet identified, lots of turtle fragments—virtually all shell material, probably from *Kallokibotion*—and a partial tooth that appears to belong to the duck-billed *Telmatosaurus*.

The quarry workers have to recross the Sibişel River farther down by means of another makeshift bridge, this time a long, narrow slab of concrete, all the while pushing a gaggle of stubborn sheep out of the way as a shepherd watches from under a tree and his dog growls territorially at us. We are making our way to the Lower Quarry, one of the sites we're presently working. Thus far, the Lower Quarry has produced teeth attributable to crocodilians, *Zalmoxes*, and a few small theropods. Right now, a *Telmatosaurus* scapulocoracoid (shoulder girdle) and two sauropod humeri are visible in the quarry. Today we're removing what looks to be a modest-sized braincase, but we cannot yet say to whom it belongs.

As the morning progresses, the temperature in the quarry rises to 38°C (100°F) or more, and the humidity reaches as much as 80%. We have no cover from the sun in the quarry, but there's modest shade to be had in a nearby scrub shelter. Quarry work is not particularly high tech. We rely mostly on tools generally found at home: chisels, paint brushes, ice picks, glue bottles, and other hand tools. There's also a stash of rock hammers for everyone; a pry bar, shovels, and a pickax for removing soil and rocks as the perimeter of the quarry becomes ever larger; and, when the work demands it, a jackhammer and a generator to blast though the more resistant overburden. We also have some acetone and polyvinyl acetate beads to make the dilute hardener for preserving our fossils. Most of the specimens, small and often isolated, are simply removed from the rocks with a modicum of effort but the greatest of care. A label is made to indicate where it was collected, and then the specimen is placed in a bag, to be stored at camp. Occasionally, a jacket made of plaster and strips of burlap is constructed around the larger and more fragile specimens to keep them intact. Meanwhile, back at camp, there are a few people whose job it is to squat in the shallows of the Sibişel River, attending to bags of cheesecloth that contain loose sediment and, we hope, small fossils from several outcrops of the Sânpetru Formation. Our wet-screening efforts are designed to find tiny dinosaur teeth or the fossil remains of fish, amphibians, lizards, or mammals, the smaller inhabitants of the region during the Late Cretaceous.

Despite the heat and incumbent discomfort (or maybe because of it), time spent prospecting or quarrying gives everyone a chance to learn about each other. With luck and some ill-defined, essential ingredients, the beginnings of a community take place. Names are learned—Cristina, Bogdan (Bob), Cosmin, Vali, George, Ovidiu, Mircea, Nicu, Roxanna, Florin—and personalities surface: the leaders, followers, clowns, fixer-uppers, and just plain hard workers. Later, during meals, rest times, and at night, the rhythm of the place transforms the crew into a family. This family fuels our dinosaur expedition, a joint, international effort between Muzeul Civilizației Dacice și Romane Deva, Romania, the museum where our fossils will later reside, and the Johns Hopkins University in Baltimore, Maryland, U.S.A.

By noon, we're in considerable need of rest and food, so it's back to camp. Lunch preparation is a traded-off duty shared by the crew. It usually consists of some combination of fresh veggies, rice, pasta, potatoes, or cabbage; meats such as mici (pronounced "meech": small, spicy grilled sausages) or salami; and bread. Pate de ficat (liver pâté) is the most appreciated food in the field. The stove runs poorly; it's American made, but forced to run with leaded fuel. As a consequence, it needs to be cleaned out daily. How it manages to run for the duration of the summer, only the deities of fieldwork can say.

How we obtain our food and water takes some planning, particularly since we don't have regular use of a vehicle or a freezer. Consequently, we bring most of our food with us from Deva when we first travel from the museum to camp at the beginning of the field season. With a population of nearly 70,000, Deva, the county seat of Hunedoara County, has a great outdoor market and quite a few grocery stores for buying our essentials. Canned fruits, vegetables, and meats are available in Hațeg, a city of 11,000 and our closest connection to the civilized world. Other necessities —such as cheeses, milk, bread, and fresh vegetables—are bought from the villagers in Sânpetru. For our supply of water, we walk up the road to the outskirts of Sânpetru, where a farmer has allowed us to fill up water jugs from his spring. Beer—Hațegana, brewed in the city of Hațeg, of course—can be had from the local bar in Sânpetru.

Similar to the surrounding villages, Sânpetru consists of about a hundred or so dwellings, two churches (Orthodox, Baptist), a store where you could buy some foodstuffs, and two bars. In many ways, a visit to Sânpetru is like stepping into the late nineteenth century. The streets are dirt roads, traveled mostly by cows, sheep, geese, and a horse or two, with an occasional car or truck as the only acknowledgment of the present. The houses, one-story buildings of brick and clay topped with brick-orange tile roofs, are all gated so that family activities do not spill out into the road. Each house has a small vegetable plot, and the villagers also farm along the floodplain of the Sibișel River. Grazing is a communal affair, where everyone heads for the fields in the morning and shares the duty of watching the animals for the day, returning when it begins to turn dark. Harvests from the field are brought into town on wagons pulled by cows or horses. One road leads to nearby Săcel, another little village where there is a small nineteenth-century castle, now used as a storage building by an orphanage. Although now run down, the castle must have been magnificent when it belonged to a family of nobles, the Nopcsa family, prior to the 1920s.

The region surrounding us is quite beautiful and peacefully rural, with vistas of hills and mountains wherever you look. This is the heart of the some of the very early recorded human history in Transylvania, that of the Dacians (pronounced "Dachians"). They were a sedentary tribal people of Indo-European descent who lived in the Transylvanian region (then known as Dacia) between the second century BCE and first century CE. Agriculture and metallurgy were their chief occupations. The wheat, fruit, wine, honey, and wax they produced were traded to the Greeks for jewelry, ceramics, sheer fabrics, and oil. From their extensive iron-mining efforts, the Dacians made weapons and a great variety of tools that were superior to the earlier bronze tools. They also crafted jewelry and vessels from the gold and silver they exploited.

The Dacians built a sophisticated and powerful defense system—a series of fortresses and fortified cities around the capital of the Dacian kingdom, Sarmizegetusa Regia—especially during the reigns of King Burebista (70–44 BCE) and King Decebal (87–106 CE). Part of this defense ring can be seen 30 km to the northeast of where we are currently working. Sarmizegetusa Regia—as well as nearby fortresses at Costești, Blidaru, and Piatra Roșie—were the last points of resistance of the Dacians during the campaign of conquest by the Roman emperor Traianus (53–117 CE) in 105 CE. After conquering Dacia, the Romans expanded the mines, especially those previously used by the Dacians to obtain gold, and also started many new ones. To maintain peace in their newly conquered region, the Roman administration and its large army required new cities and camps; by the end of their rule, they built dozens across the former Dacian kingdom. The most important was Ulpia Traiana Sarmizegetusa, the capital of Roman Dacia, which can still be seen about a dozen kilometers due west of where we have set up camp. It was large, with all the features of a Roman city—streets, aqueducts, a small arena or coliseum, a palace, and a number of temples. The Romans flourished for a long time in Dacia, no doubt traveling and carrying on commerce right here along the Sibişel, but after two centuries they had to abandon the region, due to the weakening of their empire.

Much of the history of the Haţeg region since then is preserved in its small churches. They're found in all of the villages in the area—in Sântămăria-Orlea, Pui, Săcel, Sânpetru, and Vălioara—but the Orthodox church in Densuş is the best known. A bizarre but very elegant church of some mystery—once thought to be a mausoleum for a Roman general or a temple for the god Mars—the Densuş church apparently was built in the thirteenth century, on a tenth- to eleventh-century foundation, from stones taken from the ruins of Ulpia Traiana Sarmizegetusa. The unique and somewhat strange architecture is reminiscent of the Romanesque style, but the construction was certainly cobbled together from tiles, tombstones, and columns, with a pagan altar in the center of the church, as well as containing small statues and a stone-carved figure of a lion by a local builder who seems to have had little regard for strict architectural design.

That's enough historical and geographic recounting for now; it's time to head back to the field. In the afternoon, half of the crew again heads for the Lower Quarry for more work and the other half—the prospectors —head for Târnov Hill, in the middle of the Sibişel Valley, where the Sânpetru Formation is exposed. It looks entirely covered with scrub forest, but it may hide some interesting areas of outcrop. In any event, we haven't searched it before, so off the dinosaur hunters go.

After our work in the morning, there are no fossils showing in the Lower Quarry, but we're confident that there are more bones and teeth to be had. So with jackhammer, pickaxes, and shovels, we begin enlarging the quarry. By 6:00 p.m., the quarry is now almost double its earlier size and ready for the finer-grained work of finding and removing the teeth and bones that are yet to be discovered at the Lower Quarry. So we troop back to camp, encountering on the way the outcrop wanderers. They have had some success, finding a crocodilian tooth in one place and, in another, a mish-mash of highly weathered bone scraps. We'll head back there tomorrow to see if there are better specimens preserved beneath the ground; if so, then we'll open up another quarry. But now it's time to eat. Dinner is usually like lunch—bread, butter, and jam, roasted corn, and spaghetti. However, today someone brought back from Sânpetru a homemade mămăliguța, a peasant dish made from crushed cornmeal that is also known as polenta.[1] Considered by many to be the national dish of Romania, it is certainly one of the oldest, with recipes dating back beyond Romanian history to the Etruscans of the ninth century BCE. From the Etruscans to the Romans to the Romanians to us at our paleontological field camp.

Most days it's sunny, hot, and humid in the field, but sometimes we get rain. Unlike other regions rich in dinosaur fossils, where rains are rare (though violent when they come), Transylvanian rains are much more common. Averaging 5 inches or more per month over the summer, a Transylvanian shower can last a while—throughout an afternoon, overnight, or for several days. The only thing left to do is wait it out, get cranky, realize that crankiness is an appalling way to occupy your time, and, when the rain stops, take down the tent and let it air dry in the sun. The mud-covered paths, rocks, and fossils will also need drying-out time.

Beyond an occasional rain, we also receive visitors in camp, some to help, some just passing through. A Sunday hardly ever goes by that the parents of some of the crew don't stop by with meat dishes (sarmale: stuffed cabbage leaves; mușchi de porc: pork cutlet), soups (ciorbă de burtă cu smântână: tripe soup with sour cream), and desserts (pastries like plăcintă and clătite, and even cookies)—it's really a feast on Sundays. A crew of paleontologists and natural history aficionados come from Trieste, Italy, eager to join the dig. A monsignor from a nearby Orthodox church heads on foot toward a parish who knows where. Three generations of Gypsy women, barefooted and bedazzling in their long, brightly colored skirts, wander through the woods.

During the day, but especially at night, we share our world with mice, bugs, feral dogs, and mosquitoes. In the heat of the day, we see blue-headed and little brown rock lizards. There are also rumors of vipers in the area, but they appear to have no interest in paleontologists. Bears range wild in Retezat National Park, Romania's first, located in the Retezat Mountains to the south of us, but they've never been reported to roam where we're camped. Occasionally, hawks glide effortlessly above us, probably looking for things other than fossils from their vantage point. Of all the birds of the area, we always look for wild storks nesting on chimneys or telephone poles—these birds are known to bring good luck.

At night, it again turns cool and clear. The mountains look especially beautiful, set as a shadow enshrined in an ever-darkening sky. Before it's too dim, field notes are written, someone pulls out a guitar, maybe there's a campfire. Small bats cavort in the early, comforting quiet of the night. Soon the first stars flicker, maybe a satellite, eventually to be drowned out by the outrageous spectacle of the Milky Way, best seen during an occasional rendezvous with the corner bush in the middle of the night.

This, then, is the Joint Muzeul Civilizaţiei Dacice şi Romane Deva–Johns Hopkins University Transylvanian Dinosaur Expedition.

And it all begins again tomorrow.

FRANZ BARON NOPCSA AND THE DINOSAURS OF TRANSYLVANIA

One spring day in 1895, 12-year-old Ilona Nopcsa took a walk in the hills surrounding her family's baronial estate (Plate II) outside the small village of Szacsal.[2] There she found, by chance, the first fossilized bones recognized from the Transylvanian region. Taking them home to show her 18-year-old brother, Franz Baron Nopcsa, these fossils apparently had an immediate fascination for him. Ilona showed him the site, and he later returned on his own, collecting additional dinosaur fossils before returning to Vienna to begin his university education.

When he entered the University of Vienna, Nopcsa sought out Professor Eduard Suess (figure 1.1), an eminent geologist at the University of Vienna, for advice.[3] By his own admission, Nopcsa knew nothing about paleontology, osteology, or biology, to which Suess is supposed to have said, "Then learn it." Nopcsa might have chosen any sort of career for himself, simply because of his brilliance and nobility. Yet Ilona's serendipitous discovery at Szacsal provided the spark for Franz's choice of a career, and he committed himself thereafter to the study of these fossils.

Yet how to begin? At that time, the university lacked even the most elementary educational possibilities in vertebrate paleontology. Without courses to take and textbooks to study, Nopcsa was forced along an autodidactic route. He educated himself by contacting researchers elsewhere in the world, by teaching himself comparative anatomy, and by bringing all of this information to bear on his new fossils. The material—a severely crushed skull, lower jaws, and some vertebrae—amounted to a paleontologist's worst nightmare (figure 1.2). Where did one bone stop and the adjacent bone begin? Due to its considerable deformation, how could each skull element be reconstructed? And with such obvious problems, how could these fossils be reliably placed in a taxonomic scheme with other dinosaurs known from elsewhere in the world? Despite these quandaries, Nopcsa persevered, painstakingly preparing the fossils to glean as much information as possible about his new Transylvanian dinosaur. This material was to become the type specimen for a new genus and species of dinosaur, the duckbill known as *Telmatosaurus transsylvanicus*, described by the 22-year-old Nopcsa and published in the Memoirs of the Imperial Academy of Science of Vienna in 1900.[4]

Figure 1.1. Eduard Suess (1831–1914)

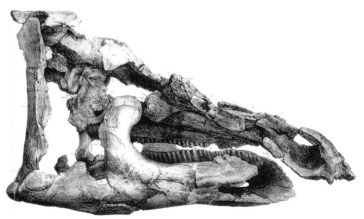

Figure 1.2. The first dinosaur known from the Haṭeg Basin: the hadrosaurid *Telmatosaurus transsylvanicus*. (Original plate from Nopcsa 1900)

Bringing It All Back Home

Nopcsa's recognition of a new form of dinosaur was no small achievement. *T. transsylvanicus* was the first dinosaur known from Transylvania, discovered at a time when very few good examples of dinosaur skulls and skeletons were known.[5] Following tradition, Nopcsa's paper began with a meticulous, if inevitably boring, anatomical description. Yet he knew that there were processes behind the formation of these bones, shaping them, providing them with the means to move against each other and to form rigid capsules around sense organs and the brain. With these things in mind, and the basic description completed, Nopcsa ventured into a less tangible, more interpretive research realm, becoming the first to attempt a reconstruction of the brain, nerves, musculature, and blood supply of a dinosaur head. This innovative perspective would soon tie Nopcsa to the central European school of paleobiology, a discipline whose goals were to link interpretations of the hard and soft anatomy of extinct organisms, their functional morphology, and their significance in evolutionary theory.[6] Nopcsa's paper was to bring him into contact with Louis Dollo (figure 1.3), the Belgian doyen of paleobiology during the last quarter of the nineteenth century, who praised the young researcher as "a comet racing across our paleontological skies, spreading but a diffuse sort of light."[7] Along with Othenio Abel (Nopcsa's contemporary at the University of Vienna), Jan Versluys, Otto Jaekel, Carl Wiman, and Ferdinand Broili, Nopcsa came to be one of paleobiology's greatest protagonists.[8] Of relevance here are Nopcsa's analyses of sexual dimorphism and growth and development in extinct vertebrates, and his soft-tissue reconstructions, as well as his studies on skin color and analyses of jaw mechanics and locomotion.[9]

Figure 1.3. Louis Dollo (1857-1931)

With *Telmatosaurus* seeding the clouds, there was now a deluge of discoveries in Transylvania. As well as 14 brief studies, Nopcsa published six longer monographs on the Late Cretaceous vertebrates from the Hațeg Basin (Plate III, top). In addition to his first monograph on *Telmatosaurus*, three other monographs focused on the ornithopod *Rhabdodon* (now known as *Zalmoxes*[10]) and were published successively in 1902, 1904, and 1922. Another, dedicated to the Transylvanian turtle *Kallokibotion*, was published in 1922. The last, on the ankylosaur *Struthiosaurus*, appeared in 1929. Nopcsa intended to write a seventh monograph on the sauropod *Magyarosaurus* with his close friend Friedrich von Huene (figure 1.4), one of Germany's most prominent vertebrate paleontologists, but this collaboration never took place.[11]

NOPCSA AND ALBANIA

Although we regard Nopcsa's studies of the Transylvanian dinosaurs as his greatest legacy, he is equally well known for his work on Albania.[12] Beginning at the turn of the twentieth century, Nopcsa began a short but intense relationship with the remote Balkan country of Albania.[13] He had first been introduced to the Albanian region in 1899, when he met

Figure 1.4. Franz Baron Nopcsa (*left*) and Friedrich von Huene (1875–1969; *right*) inspecting the fossil-bearing, Lower Jurassic strata at Holzmaden, Germany. (Photo courtesy of Universität Tübingen, Germany)

Ludwig Graf Drasković, a lieutenant in the Austro-Hungarian Army who was returning from this part of the Balkans. Apparently it was Drasković who overwhelmed Nopcsa with thrilling stories of rugged mountains, poor but proud villagers, and blood feuds. It was not until after his graduation from the University of Vienna, however, that Nopcsa made his first trip to Albania, thanks to a graduation gift of 2,000 Austrian crowns from his uncle and namesake, Franz Baron Nopcsa (Franz Baron Nopcsa, the elder, was the First Lord Chamberlain to Elisabeth, Empress of Austria and Queen of Hungary). Traveling first through Greece in 1903, then through the eastern Balkans, Nopcsa finally made his way to Skutari (now Shkodra) in northern Albania, where he lived with a local family. His extensive travels in the mountains of northern Albania over the next 15 years provided the basis for his detailed accounts of the geology and geography of the region, which he used later in his scientific career in support of Wegener's theory of continental drift. It also gave Nopcsa the opportunity to learn about the laws, customs, and people of this remote area of the ever-explosive Balkan Peninsula. Living with and working alongside members of the Merdite tribe, Nopcsa integrated himself into the community, amassing considerable information about the tribes of northern Albania: their history, languages, and religious practices. His major works—amounting to well over a thousand pages of text—are still considered among the most significant in Albanology.

The ethnic and political problems of today's Balkans are deeply rooted in history, manifested, for instance, in the Balkan Wars of 1912 and 1913. At that time, the Austro-Hungarian and Ottoman empires politically dominated the region, with tensions in the northern Balkan area being especially high. No one could have appreciated this situation more than Nopcsa. He had lived with the people, learned their language and dialects, been involved in a blood feud, and was a leader of soldiers. As a stalwart supporter of Austro-Hungary, he feared Turkish and Serbian aggressions from the south and east, and sought solutions through his political connections and influence. He pushed for an independent Albania allied to Austro-Hungary, a union presumably desired because of his love for his adopted Albania and his loyalty to the Empire. He used whatever influence he could muster to sway Count Pál Teleki von Szék and Count István von Bethlen, two fellow Transylvanians destined to be

Figure 1.5. Nopcsa's secretary, Bajazid Elmas Doda (1888–1933; *left*), and Franz Baron Nopcsa (*right*). (Photo courtesy of the Hungarian Natural History Museum, Budapest)

successive prime ministers of Hungary, and the general chief of staff of the Austro-Hungarian Empire, Count Franz Conrad von Hötzendorff. Nopcsa's plan was to arm the northern Albanian tribes and carry out a guerrilla war against the Turks, routing them and liberating Albania. Nopcsa also had his name put forward as a potential candidate for King of Albania in 1913, should his campaign have resulted in success. It was Prince Wilhelm zu Wied, however, who got the job, even though the latter held it for only six months before being driven out by the Albanians. The 1914 murder of Archduke Franz Ferdinand in Sarajevo was a turning point for Nopcsa; although he continued to publish his often-massive Albanian studies until his death, he was never to return to his adopted country after 1916. His involvement in the geopolitics of the Balkans did not stop with Albania, however. During World War I and immediately afterwards, he apparently also carried out espionage in western Romania under the auspices of the prime minister of Hungary, Count István Tisza.

It's a pity that Nopcsa was never able to fully explore the paleobiology, paleoecology, and evolutionary dynamics of the Hațeg fauna. For on 25 April 1933, Nopcsa's body and that of his Albanian secretary, Bajazid Elmas Doda (figure 1.5, on previous page), were found by police at their Singerstrasse residence in Vienna. A note at the scene, written in Nopcsa's hand, made clear the last moments of the lives of these two men: Nopcsa had doctored Bajazid's morning tea with sleeping powder and then had shot him; thereafter Nopcsa had put a gun to his own head to commit his final act.

CHAPTER 2

Dinosauria of Transylvania

At the time of Franz Nopcsa's death in 1933, the Hațeg fauna was thought to include five dinosaurs, a bird, a crocodile, a turtle, and a pterosaur. Unfortunately, work in Transylvania went fallow after the First World War, when the defeated Austro-Hungarian Empire ceded Transylvania to Romania. It was not until the mid-1970s that the collection of vertebrate fossils resumed in the Hațeg Basin. In 1978, two teams came together to follow in Nopcsa's footsteps. One was supervised by Dan Grigorescu from Universitatea din București, and the other was originally organized by Ioan Groza, later supervised by one of us (Jianu) and then by us both, under the auspices of Muzeul Civilizației Dacice și Romane Deva.[1] These two groups, plus a more recent joint expedition from Universitatea Babeș-Bolyai Cluj Napoca, Romania, and the Institut Royal des Sciences Naturelles de Belgique in Brussels (under the supervision of Vlad Codrea and Thierry Smith, respectively[2]) have ventured to the outcrops at Vălioara, Densuș, and the Sibișel Valley, as well as several new locations, this time in the Transylvania Depression—Jibou (Sălaj County) in the north and Oarda de Jos, Vurpăr, Bărăbanț, Lancrăm, Sebeș, and Vințu de Jos in Alba County, along the southeastern margin of the Apuseni Mountains north and east of the Hațeg Basin.[3] These efforts have amassed several thousand new fossil specimens and added considerably to the diversity of the known assemblages. Several different kinds of dinosaurs are represented for the first time, as well as new bony fish, amphibians, mammals, lizards, and crocodilians. Along with this richer picture of the fauna has come a better understanding of the paleoecological context of the Hațeg Basin and other localities, as well as the evolutionary significance of this part of the world during the Late Cretaceous.[4]

The best-known members of the Hațeg and similar assemblages clearly are the dinosaurs. Nopcsa knew or named nearly all of them, including *Telmatosaurus transylvanicus* and *Zalmoxes robustus* among ornithopods, the armored *Struthiosaurus transylvanicus*, the sauropod *Magyarosaurus dacus*, and a theropod he referred to *Megalosaurus*, a poorly known form first identified from the Middle Jurassic of England.[5] Given the Late Cretaceous age of these deposits, it is likely that many of these dinosaurs were among the last of their dynasties.

In this chapter, we hope to accomplish two things. First, we want to put members of the Transylvanian menagerie into their evolutionary or phylogenetic context. Second, we want to breathe life into the fragments of ancient bones from Transylvania and, thereby, get a meaningful picture of these beasts that once roamed the Transylvanian region. We begin by outlining a field of study called phylogenetic systematics, otherwise known as cladistics. Cladistics is used to establish who is more closely related to whom among a group of organisms. We also use it to understand the relationships of the dinosaurs and their Transylvanian cohorts along the way.

THE HISTORY OF LIFE AND HOW WE KNOW IT

Fossils are the petrified remains of prehistoric life, something that has been recognized in the scientific community for three centuries, ever since the Danish scientist Nicholas Steno (1638–1686) first interpreted fossils as the vestiges of once-living creatures. Darwin understood that the links among these organisms constituted evolution, and he postulated a mechanism for the latter that depended not on divine design, but on the day-to-day action of environment on variable individuals within a population. He also understood that evolution, by its continual production of generations of descendants from earlier descendants, from still earlier descendants, and so on, back to primordial ancestry—in other words, diversification—was therefore hierarchical. It's not for nothing that his canon—"descent with modification"—emphasized the hierarchical property of evolution. Others had already recognized this hierarchy in nature, most notably Carolus Linnaeus, the great eighteenth-century classifier of all organisms. His cataloging revealed that God's creative hand had hierarchical tendencies, and all organismal taxonomies have had this structure thereafter. Darwinians and everyone since then have taken on Linnaeus's practice of, if not his motivation for, identifying hierarchies in nature, because these nested sets of diversity conform to a single phylogeny, a single genealogy into deep time that documents the interrelatedness or connectedness of all life.

How, then, can we identify phylogeny's hierarchy? We don't have a written record, as we do for our personal family genealogies, but we do have the similarity of features possessed by organisms that underlie their evolutionary history. From Darwin's time to the present, we have given special evolutionary significance to a particular class of similar features, called homologies, whose presence gives nature its hierarchical property. For a feature to be a homology, it must have evolved only once. How we determine whether a feature has a unique origin is a matter of comparison. Take, for example, the presence of retractile claws in mammals. Such claws are very anatomically and biomechanically similar to each other in all the mammals that possess them—cats, lions, tigers, and others. Here we've passed the first test of homology, the test of similarity. The second test is the congruence of a feature with phylogeny (i.e., with evolutionary history): can we tell if retractile claws evolved just once, instead of several times, in the groups of animals that have them? In our example, the answer is yes; all cats have retractile claws and, because we regard all of these animals as very closely related, we hypothesize that retractile claws evolved once in the group of mammals called Felidae.[6] Said another way, this homology evolved once in the common ancestor of Felidae and then was passed on to its descendants. By the same token, if someone tells us that a given mammal is a felid, we would expect it to have had retractile claws, although they may have been subsequently lost. By their very nature, homologies evolve only once, and they therefore speak about the closeness of the relationship between two kinds of organisms.

By properly identifying a group of organisms as having a single common ancestor (a single origin) on the basis of its characters, we have begun determining whether that group is monophyletic. The other important aspect of monophyly is that the group contains all the descendants of this common ancestor. Seen in this way, *Homo sapiens* is monophyletic, felids are monophyletic, birds (Aves) are monophyletic, and Dinosauria is monophyletic. But Dinosauria without birds is not monophyletic—it leaves off some of the descendants of the common ancestor of all dinosaurs. This latter kind of grouping, known as paraphyly, is similar to leaving off Uncle Bob and his family from your family tree—you may want to, but the end result wouldn't be a true reflection of your family history.

The interdependence of homology and hierarchy forms the basis of what is known as cladistic analysis, a tool that is particularly well suited to reconstructing phylogeny. Cladistic methods, first developed by the German entomologist and founding father of phylogenetic systematics, Willi Hennig (figure 2.1), seek to establish the hierarchical nature of evolution by searching for the nested arrangement of organisms and the features they possess.[7] This pattern is then portrayed on a branching diagram called a cladogram, with each collection of branches being referred to as a clade. The characters used to justify the branching pattern in a cladogram may have broad or limited distributions, such that some characters will diagnose more general relationships, while others will diagnose more restricted ones (figure 2.2). For example, Aves (modern birds) and Crocodylia are both diagnosed as archosaurs because of their braincase and palate, general features that evolved once in the common ancestor of these two groups, although several aspects of these features may have later altered within the groups. More specific affinities, for example *Homo sapiens* and *Australopithecus afarensis* within Hominidae, can be assessed by identifying homologies—both pos-

Figure 2.1. The founder of phylogenetic systematics (also known as cladistics), Willi Hennig (1913–1976)

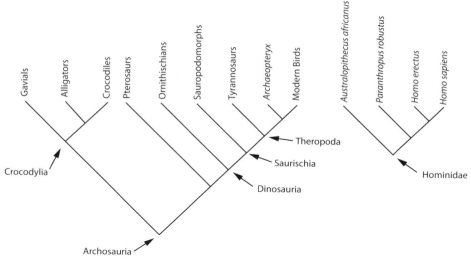

Figure 2.2. Cladogram of Archosauria (*left*) and Hominidae (*right*)

sess features of the pelvis, femur, and knee that relate to their unique evolution of bipedality. In other words, a character may be specific to one group (i.e., bipedality in Hominidae), but general in a smaller subset of that group, because it is now being applied at a different position in the hierarchy.

If things were as simple as this, the task of determining phylogeny would be a snap. We would seek the nested relationships of similarities in molecules, morphology, and behavior, then interpret them as homologies, and the hierarchical history of life would be revealed. All similarities would be homologies, from which we could then reconstruct phylogeny.[8] However, nature, like history, is messy—she invents similarities that are not and cannot be homologies. Take, for example, the presence of eyes in both dolphins and squid. They are not considered as having evolved just once in the common ancestor of Vertebrata and Cephalopoda; given the other features and taxa that must be considered in this comparison, eyes must have evolved at least twice to account for their distribution. Said another way, these nonhomologous similarities, called homoplasies, are not congruent with a single origin during phylogeny. Homoplasies are those features that get in the way of discovering phylogenetic patterns, because they are produced by two or more events in evolutionary history.

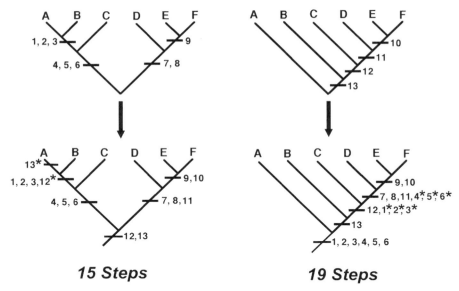

Figure 2.3. A hypothetical example of how to determine the most parsimonious cladogram (see text for explanation)

Let's take a final look at how characters can be identified as homologies and homoplasies, using a hypothetical example. We have six taxa (A–F) as well as two phylogenies of these taxa that we want to test, to see which one is best supported by the features we've hypothesized as being homologies. In figure 2.3, the cladogram on the upper left consists of two monophyletic groups (A–C and D–F), the first supported by three homologies (characters 4–6) and the second by two (characters 7–8). Within each of these groups there is another monophyletic clade, each supported by its own homologies (characters 1–3 for the first group and character 9 for the second).

The cladogram in the upper right implies dramatically different relationships from that in the upper left. All taxa, from A to E, are sequentially more closely related to taxon F. These relationships are supported at each level by single homologies (characters 10, 11, 12, and 13), each of which is different from those that produced the cladogram to its left. We can tell which tree is likely to be correct, given the data we have, by finding out which tree is shorter. In doing so, we must use not only the data that support a given cladogram, but also all the others that don't support it. That is, we must also account for all the characters on each tree (i.e., by applying characters 10–13 to the left tree, and characters 1–9 to the right tree). We've done that for the cladograms in the lower row.

Because *all* characters must appear on the cladogram, conflict among features is bound to occur. These conflicts in cladograms and their character support are settled by letting the possible homologies fight it out among themselves. This rather bloodless battle is carried out in the arena of parsimony, otherwise known as Ockham's Razor. Parsimony, originally articulated by the fourteenth-century Franciscan monk William of Ockham (in England), ensures that, in the case of cladistic analyses, the simplest or shortest cladogram is chosen. In following this principle, we therefore have to select the tree supported by the greatest number of homologies from those that we originally hypothesized. This is a requirement, since homologies, the *sine qua non* of close relationships, are represented only once on a cladogram, but homoplasies, by their very nature, are found multiple times on the cladogram. In other words, some of our original hypothesized homologies are forced to become homoplasies by the weight of evidence coming from the other characters. Depending on the situation, some characters will either evolve twice or evolve and then reverse back to their more primitive condition (shown with an asterisk). Once the distribution of all the characters is determined, their positions on each cladogram are counted up. The shortest tree, which we call the most parsimonious one because it maximizes the number of homologies, is the one on the left, with a length of 15 steps. It is four steps shorter than the one on the right, which takes 19 steps to account for all the evolutionary changes in the six taxa.

THE SCIENCE (AND ART) OF COMPARATIVE ANATOMY

To understand the biology of extinct creatures (such as the denizens of ancient Transylvania) as best we can, it behooves us to do what every paleontologist must—fill in the missing bits of skeleton and teeth in a reasonable way to better understand the no-longer-living creatures we're studying. We owe the means by which we meet this challenge to the founding father of dinosaur paleontology, Sir Richard Owen (figure 2.4), in his role as a comparative anatomist.[9] Owen's greatest achievement in this arena has become one of the most cited and mythologized of all episodes in paleontological history. In 1839, Sir Richard famously

Figure 2.4. Richard Owen (1804–1892)

deduced the existence of an enormous flightless bird that once inhabited New Zealand, reconstructing the whole animal from a single fragment of a femur. The test of his construction—and the basis for his eminence as a comparative anatomist—came with the discovery, four years later, of well-preserved skeletal remains of the extinct moa, *Dinornis maximus*, a 3.5 m tall, ground-dwelling, flightless bird.[10] Based not on arcane knowledge, witchcraft, manipulative card tricks, chicanery, or wild speculation, Owen's achievement instead was due to years of studying the structure of a wide array of living and extinct animals, or what we now call comparative anatomy.

Like Owen, but also with insights that come from cladistics, we will use comparative anatomy to develop a better understanding of the Transylvanian dinosaurs, most of which are known from tantalizing and abundant, but fragmentary, material. As comparative anatomists, we will look for similarity in the sizes, shapes, and features of bones; the form and density of serrations on teeth; and anything and everything else that we can to better understand the creatures themselves. Of course, we can only infer so much from these features, but at least they're something tangible with which to begin our assessment of the Transylvanian dinosaurs and their evolutionary legacy.

DINOSAURS

One last thing before tackling the Late Cretaceous inhabitants of Romania: what exactly are dinosaurs? The term Dinosauria, meaning "fearsomely great reptiles," was coined a century and a half ago by Richard Owen for the agglomeration of what was then the poorly known, gigantic, prehistoric reptiles from the Mesozoic Era.[11] Thanks to Harry Govier Seeley (figure 2.5), toward the end of the nineteenth century we have recognized that all dinosaurs are either saurischians (lizard-hipped) or ornithischians (bird-hipped).[12] Said another way, Dinosauria (figure 2.6) is a monophyletic group, composed of Saurischia and Ornithischia and their common ancestor. We have known about each of these groups for as long as we have known about dinosaurs; the saurischian *Megalosaurus* and the ornithischian *Iguanodon* were discovered in England about 1820. Saurischia includes both the gigantic sauropods and the carnivorous theropods, familiar to museum- and movie-goers everywhere as the long-necked *Apatosaurus* and the fearsomely toothed *Tyrannosaurus rex*, respectively. The other major dinosaur group, Ornithischia, comprises a vast array of plant eaters that includes duck-billed, armored, and horned dinosaurs (such as *Triceratops*).

Figure 2.5. Harry Govier Seeley (1839-1909)

In all saurischian dinosaurs, the pubic bone slants down and forward. This is hardly a unique feature; so do the hips of many other extinct and living animals, including *Homo sapiens*. In fact, lizard hips are so common among land-living vertebrates that paleontologists must look to other features uniquely shared by these dinosaurs in order to recognize them as real, or monophyletic, evolutionary groups. Fortunately, several other characters are commonly used to unite this group (figure 2.7), including a hand with a large thumb and elongate second digit, and elongation of the neck, among other features.[13] The various subgroupings of saurischian dinosaurs, the most important of which (for our narrative) are sauropods and theropods, coalesce on the basis of many additional features.

Ornithischians (figure 2.8), a diverse group of plant eaters, are known as the bird-hipped dinosaurs because their pubic bone is rotated backward, much like a bird's, although the two conditions do not have the

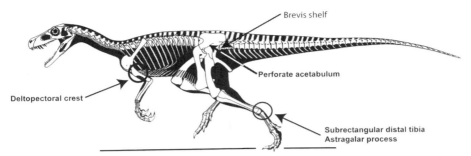

Figure 2.6. The skeleton of *Herrerasaurus*, indicating several of the unique features of Dinosauria: an elongate deltopectoral crest on the humerus; a brevis shelf on the postacetabular part of the ilium; an extensively perforated acetabulum; the tibia with a transversely expanded, subrectangular distal end; and the ascending astragalar process on the front surface of the tibia

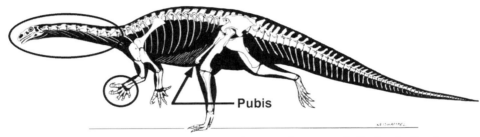

Figure 2.7. The skeleton of the prosauropod dinosaur *Plateosaurus*, indicating the lizard-hipped pelvis of saurischian dinosaurs, as well as several unique features that unite Saurischia: elongation of the neck, and a hand with a large thumb and an elongate second digit

same origin and therefore are not homologous. One explanation advanced for this characteristic is that it evolved to provide more abdominal room for the fermentation of the plants they ate. Ornithischians share a number of other unique features, including a special bone (called the predentary) covering the tip of the lower jaws that gave extra strength to this region during biting.[14] In addition, the back teeth were low, triangular, and well designed for chewing. Because these teeth were positioned away from the outer margins of the jaws, a space between them and the side of the face may have been enclosed by muscular cheeks to prevent food from falling out of the mouth while chewing. Finally, bundles of elongate ossified tendons formed a network along most of the vertebral column, providing additional support for the body, which was

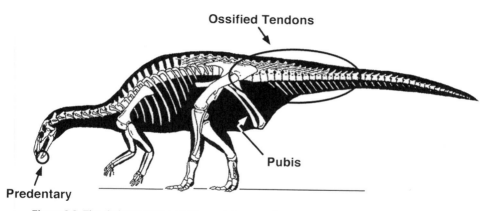

Figure 2.8. The skeleton of the ornithopod dinosaur *Camptosaurus*, indicating several of the unique features that unite Ornithischia: the predentary covering on the tip of the lower jaw, the bird-hipped condition in which the pubic bone is rotated backward, and the presence of ossified tendons along the backbone

balanced at the hip joint during a bipedal stance. Ornithischian dinosaurs come in a variety of shapes and sizes, but most can be grouped as stegosaurs, ankylosaurs, pachycephalosaurs, ceratopsians, and ornithopods. Of these, only ankylosaurs and ornithopods have thus far been recovered from Transylvanian deposits.

The Titanosaur Sauropods

Sauropods of the Late Cretaceous of Europe are best known from the present-day French Pyrenees and northern Spain, particularly through the recent excavations led by Jean Le Loeuff, Eric Buffetaut, and other researchers at the Musée des Dinosaures in Espéraza, France,[15] and field research at Laño in Basque County, conducted by Xabier Pereda-Suberbiola, José L. Sanz, and other French, Spanish, and Basque scientists. Less well known are the sauropods from the fossil beds farther to the east, thus far recognized only from the Haţeg Basin and the Transylvanian Depression in Romania. Nopcsa first reported finding sauropod bones here in 1902.[16] Consisting almost entirely of isolated postcranial skeletal elements (mostly vertebrae and a few limb bones) of relatively small but presumably adult individuals (figure 2.9), Nopcsa referred these to *Titanosaurus*, a form otherwise known at that time only from India, Argentina, Madagascar, France, and England.[17] Huene restudied Nopcsa's sauropod material, renaming it *Magyarosaurus* and recogniz-

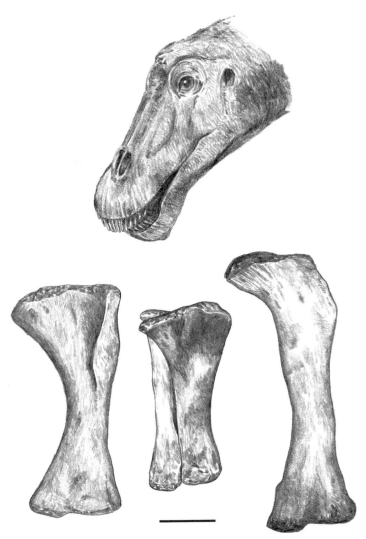

Figure 2.9. Magyarosaurus dacus: reconstructions of the head (*above*), the left humerus (*left below*), the radius and ulna (*middle below*), and the femur (*right below*). Scale = 10 cm

ing three species: *M. dacus*, *M. transsylvanicus*, and *M. hungaricus*.[18] Recent research by Zoltán Csiki and others recognizes two species from Transylvania, the smaller and more numerous *M. dacus* and a larger, rarer titanosaur, *Paludititan nalatzensis*.[19]

From the outset, all of the sauropod material from Transylvania was considered titanosaurian in nature (box 2.1). The first-named of these titanosaurs, *Titanosaurus indicus* (described in 1877, although the taxon *Titanosaurus* is now a *nomen dubium*, or "doubtful name"),[20] provided only a partial portrait of these giants, but later discoveries have helped refine our understanding of them.[21] We now know that titanosaurs had arisen by the Late Jurassic and remained common in parts of the landmasses of the Southern Hemisphere (known as the supercontinent Gondwana) up through the end of the Cretaceous. Titanosaurs have been found in the southwestern United States, Europe, and Asia,[22] but the best-known titanosaurs come from the Upper Cretaceous of Argentina and Madagascar (*Saltasaurus loricatus* and *Rapetosaurus krausei*, respectively).[23]

As is generally true of nearly all sauropods, the fossil record of titanosaurs consists mainly of disarticulated, dissociated, and often isolated material. Only a few relatively complete skeletons or skulls of any titanosaur are known.[24] Despite this imperfect knowledge of the skeletal record, we can reasonably state that titanosaurs ranged from 7 to 30 m in length. However, *M. dacus* was probably no more than 5 to 6 m long as an adult,[25] whereas the larger *Paludititan nalaczensis* from Hațeg may have been 12 to 14 m long.[26] Similar to their other sauropod kin, titanosaurs must have looked like a cross between today's elephant and giraffe, with a small head atop a long neck and a tail extended behind a rotund body supported on four sturdy legs (Plate III, bottom). Titanosaurs were unique among sauropods in one regard: their backs were covered in a pavement of bony armor.[27]

The skull of either of the Transylvanian titanosaurs, which would provide data critical to understanding their relationships and paleobiology, is all but unknown. We have no teeth, no jaws, and no facial skeleton from them. One small sauropod braincase was collected by the University of Bucharest field party from the Pui fossil locality (figure 2.10). It's a relatively tall, boxy-looking affair, with plenty of openings for the cranial nerves used in the senses of smell, sight, hearing, balance, taste, and

BOX 2.1 Evolutionary Relationships in Sauropoda

Sauropods, known to science from as early as 1841, are now represented by more than 100 species that range in age from the Late Triassic through to the end of the Cretaceous. These giants have been the most resistant to phylogenetic analysis among major dinosaur taxa, almost certainly due to the immense size of their remains (making them very difficult to work with) and the often-fragmentary skeletal remains of many sauropod taxa. Nevertheless, two researchers—Paul Upchurch from University College London and Jeff Wilson from the University of Michigan—separately took on the task of putting these giants into their phylogenetic context. There is a great degree of agreement between their work, including recognition of Neosauropoda, Macronaria, and Titanosauria as taxa nested within Sauropoda.

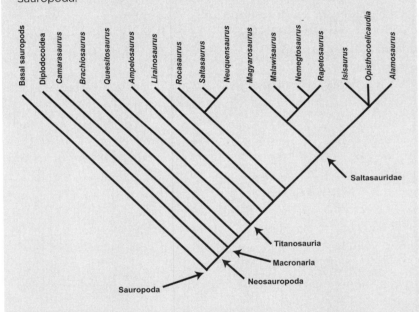

Note: See Upchurch 1995, 1998; Upchurch et al. 2004; Wilson 2002; Wilson and Sereno 1998.

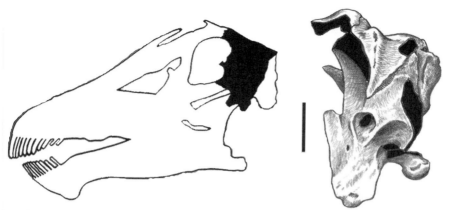

Figure 2.10. The braincase (*right*) of a juvenile titanosaur (possibly *Magyarosaurus*) from Pui, with a silhouette (*left*) indicating its position in the skull. Scale = 10 cm

other sensory and motor functions. Through a large central hole in the back of the braincase (foramen magnum), the great rope of neurons called the spinal cord traveled the length of the animal to the tip of the tail, sending nerve impulses, for example, from the brain to the muscles and organs of the body and bringing back sensations from the skin to the brain about how warm or cold the weather was. The braincase is short and extremely deep. The bones are not strongly fused together, indicating that this specimen probably came from an immature individual. Perhaps the most interesting aspect of this braincase is the two short, oval swellings on the skull roof, which look vaguely like the bases of the antlers of deer. In addition, the specimen indicates that the openings of the nasal cavity were positioned high on the face, probably in front of the eyes, much like the arrangement seen in a number of other sauropod groups. Even though these openings are high on the skull, it is now thought that the nostrils themselves were to be found toward the front of the head.[28]

To piece together other aspects of the biology of the Transylvanian titanosaurs, we must infer such characters from their close relatives.[29] The long, chisel-shaped teeth of these sauropod relatives were restricted to the front portion of the jaws, with chewing limited to an up-and-down or perhaps a front-and-rearward movement of the jaws. Tooth morphology, and especially wear, indicate that these gigantic herbivores either nipped or stripped foliage, and it is doubtful that many of these animals fed selectively, given the construction of the jaws, the nature of the dentition, and body size. Instead, titanosaurs may have ground their food using gastroliths in their muscular gizzards. Whether *Magyarosaurus* and other titanosaurs browsed at high levels within the canopy or foraged a few meters above the ground is a matter of debate.[30] Nearly all sauropods are now thought to have held their long necks nearly horizontal, except perhaps when occasionally feeding on high-growing foliage.

Despite having the smallest brains (relative to body size) of all dinosaurs, sauropod trackways indicate that these gigantic animals engaged in different kinds of social behavior, particularly gregariousness and probably migration (at least during part of the year). Some scientists have argued that it could not have been otherwise: a herd of sauropods would soon deplete its food sources in one area and then would have to move on to another in order to survive. These same trackways suggest that a migratory sauropod walked at rates of 20 to 40 km per day, although they may have been able to reach speeds of 20–30 km/hour for short bursts.

Defense in sauropods is obvious: large size confers the greatest deterrent against an attack by a predator, although not for the young, the very old, and the infirm. As for *Magyarosaurus dacus*, who these predators were is more than a little problematic.

The Transylvanian Theropods

As their teeth and powerful jaws attest, theropods were the supreme predators of the Mesozoic. They included the likes of *Tyrannosaurus*, *Carcharodontosaurus*, and *Giganotosaurus*, all vying, at up to 13 m long, for the distinction of being the largest of all terrestrial carnivores (figure 2.11). Theropods were also among the smallest—though no less dangerous—of the dinosaurs: *Velociraptor*, *Troodon*, and *Compsognathus* were all less than 2 m long. Through the evolutionary connections between *Deinonychus* and *Archaeopteryx* (box 2.2), these creatures of the past have given us living birds.[31]

Late Cretaceous theropods are best known from fossils recovered in North America, Asia, Africa, and South America; those from Europe—consisting mostly of isolated teeth, vertebrae, and various limb bones[32]—

BOX 2.2 Evolutionary Relationships in Theropoda

The study of theropod relationships is one of the most contentious and stimulating of all dinosaur research being conducted today. In part, this is because of their evolutionary relationships with birds, but it is also due to incredible new discoveries of theropods: both large and small, feathered and naked, brooding and hatching. Wonderful and exciting though these times are, they have also lent some instability to our understanding of theropod evolutionary history. To facilitate discussion, we will speak of particular theropods along the following lines. Near its base, Theropoda is split into Coelophysoidea, Ceratosauria, and then Tetanurae, the last representing the theropod line leading to birds. Within Tetanurae, we find a number of primitive forms (including *Megalosaurus*, the first dinosaur to be discovered) standing outside a group of theropods called Coelurosauria. Within this latter group, we have a number of basal forms—such as *Compsognathus*, *Ornitholestes*, and tyrannosauroids—before we come to Maniraptoriformes. This latter group includes the ostrich-mimicking theropods (Ornithomimosauria) and maniraptorans themselves. With Maniraptora, we have come very close to the origin of birds. Here we are confronted with the likes of the nightmarish oviraptorosaurs, the bizarre therizinosauroids, and, through troodontids and dromaeosaurids, to *Archaeopteryx*, emus, parrots, and the rest of Aves.

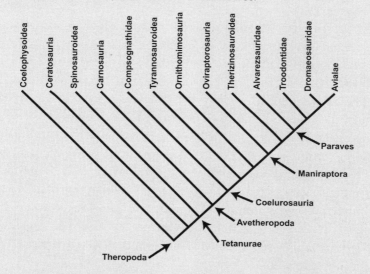

Note: See Clark et al. 1999; Holtz and Osmólska 2004; Ji et al. 1998; Norell et al. 1994, 1995; Xu et al. 2000, 2001, 2003.

Figure 2.11. The large and the small among predatory dinosaurs: reconstructions of *Giganotosaurus* (*above*) and *Velociraptor* (*below*), with a human included for scale. (After Coria and Salgado 1995)

offer intriguing glimpses into theropod diversity, but their features do not pinpoint species relationships very well. Theropods represent the most diverse component of the vertebrate biota of the Late Cretaceous of Transylvania, but elusively so—their remains, known from a meager dozen or so limb elements and teeth, are the most poorly known of the entire fauna. Consequently, paleontologists have juggled and reassigned fossil fragments to a number of different kinds of theropods. Some of these fragmentary Transylvanian fossils, as well as new collections of theropod remains, are only now beginning to be sorted out.

Nopcsa described the first "conventional" (i.e., nonavian) theropod—*Megalosaurus hungaricus*—from the Late Cretaceous of Transylvania, although the small isolated teeth on which it was based came not from the Hațeg Basin, but from older (possibly Santonian[33]) coal outcrops in the Borod Basin, some 150 km to the northwest.[34] Nopcsa attributed these teeth to *Megalosaurus* with some trepidation, fully aware that in many other species assignments to this genus, all that was usually meant was "moderately large theropod from Europe." Similar to most of these uncritical determinations, *Megalosaurus hungaricus* is now regarded as an indeterminate theropod—one, to make matters worse, whose original remains unfortunately are now lost.

The first theropod from the Hațeg Basin was originally thought to be a bird, rather than one of the more conventional theropods, such as *Megalosaurus* or *Allosaurus*. *Elopteryx nopcsai*, described in 1913 by

Charles W. Andrews, an ornithologist at the British Museum (Natural History) in London, was based on the top end of a femur and the lower end of a tibiotarsus (the product of fusion between the shin bone—the tibia—and the upper ankle bones, a feature universally found in birds, but also, as we know today, in some nonavian theropod dinosaurs). Andrews originally identified *E. nopcsai* as a Late Cretaceous pelican.[35] Andrews's interpretation was seconded by Kálmán Lambrecht, a friend of Nopcsa and himself a renowned paleo-ornithologist. He also referred three new tibiotarsi to *Elopteryx*.[36] In 1975, two researchers at the British Museum (Natural History), ornithologist Colin J. O. Harrison and paleontologist Cyril A. Walker, reexamined the available specimens of *E. nopcsai* and determined that they really came from three different forms. One was Andrews's original *Elopteryx*, but they also allotted one of the tibiotarsi to another form called *Bradycneme draculae* and another two to a second form, *Heptasteornis andrewsi*, both classified as owls (figure 2.12).[37] *Elopteryx* returned to the stage in 1981 with newly discovered material, the lower end of a femur that was described by Dan Grigorescu from the University of Bucharest and Eugen Kessler from Muzeul Țării Crișurilor Oradea (Secția de Științe ale Naturii) in Oradea, Romania; these two paleontologists argued again that *Elopteryx* was a pelican.[38] Regardless, this new specimen turns out to belong to a juvenile *Telmatosaurus* and thus doesn't help in determining the affinities of the other theropod fragments.[39]

Although they never garnered much attention in the studies of paleo-ornithologists, nevertheless things were changing for *Elopteryx*, *Bradycneme*, and *Heptasteornis*. W. Pierce Brodkorb, of the University of Florida, was the first to cast doubt on the avian affinities of these forms in 1978, suggesting instead that they represented small nonavian theropods of uncertain affinity.[40] Thereafter, many paleontologists followed Brodkorb's lead in excluding these three Hațeg forms from Aves, suggesting instead that they were troodontids or dromaeosaurids, based on their size and the age of the beds from which they were recovered.[41] We will question whether these assignments make sense, but first we will recount some exciting new theropod discoveries made in the 1990s.

The first time a portion of a theropod skull, a small one, was unearthed from a rock outcrop in the Sibișel Valley was in the early 1990s.[42] The two elements of the skull roof (figure 2.13) suggested that we had

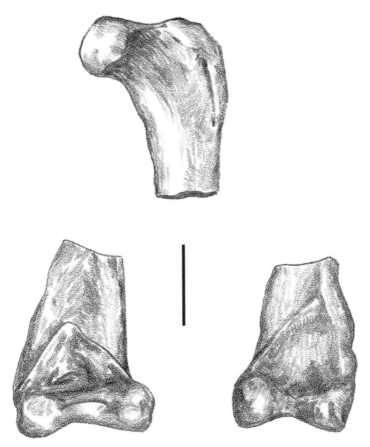

Figure 2.12. The top end of the femur of *Elopteryx nopcsai* (*top*), the bottom end of the tibia of *Bradycneme draculae* (*bottom left*), and the bottom end of the tibia of *Heptasteornis andrewsi* (*bottom right*). Scale = 2 cm

something new among the Transylvanian dinosaurs. So we compared our new find with other theropods, proposing that the new Haţeg skull fragments came from a dromaeosaurid (we ended up calling it the "Romanian Raptor"), with its greatest similarity to *Saurornitholestes langstoni*, known from the Late Cretaceous of North America.[43]

A more recent analysis of all of this theropod material—as well as of abundant, newly collected specimens by Zoltán Csiki and Dan Grigorescu,[44] and by Vlad Codrea, Thierry Smith, and coworkers[45]—has

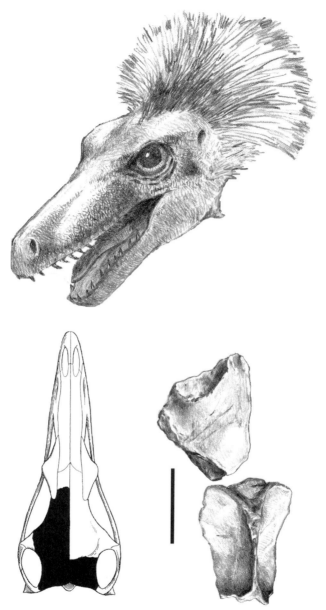

Figure 2.13. A reconstruction of the head of the Haţeg dromaeosaurid (*top*; restored after *Dromaeosaurus*), and a dromaeosaurid skull roof in dorsal view (*bottom right*), with a silhouette (*bottom left*) indicating its position in the skull. Scale = 3 cm

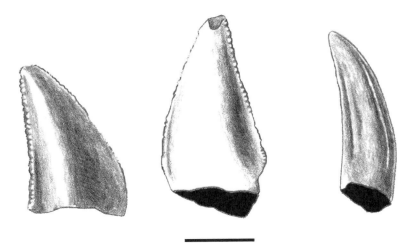

Figure 2.14. Theropod teeth from Transylvania: from a dromaeosaurid (*left*), a troodontid (*middle*), and *Euronychodon* (*right*). Scale = 3 mm. (After Csiki and Grigorescu 1998)

advanced Romanian dinosaur paleontology in two ways. First is an improved understanding of the fauna through the study of the new microfaunal collections—small teeth and bones—from several sites in the Haţeg Basin. From these, the Late Cretaceous fauna includes dromaeosaurids, a small troodontid theropod, and several curious small forms that are compared with the enigmatic *Euronychodon*, *Richardoestesia*, and *Paronychodon*, all known from elsewhere in Europe and North America at the end of the Cretaceous (figure 2.14). Second is a reevaluation of *Elopteryx*, *Bradycneme*, and *Heptasteornis* and their evolutionary relationships with other theropods. Although none of the Haţeg forms appears to be diagnostic at the species or generic level, *Elopteryx nopcsai* possesses features that suggest that its affinity lies within Maniraptora, the large theropod group that includes *Deinonychus*, *Velociraptor*, and birds. *Elopteryx* is clearly not a Late Cretaceous pelican, but a maniraptoran (possibly a dromaeosaurid,[46] such as *Velociraptor* and *Deinonychus*, or a troodontid,[47] like *Saurornithoides*). Nor are *Bradycneme draculae* and *Heptasteornis andrewsi* Late Cretaceous owls. Instead, *B. draculae* appears to be an indeterminate maniraptoran, and *H. andrewsi* is thought to be an indeterminate alvarezsaurid.

Thus far absent from the Haţeg assemblage is any evidence of a large dinosaurian predator, commonly at the top of the food chain in nearly all

Mesozoic terrestrial fauna.[48] The closest to this pinnacle is a single dorsal vertebra (5 cm in diameter) with pleurocoels, which represents no more than a medium-sized theropod.[49] Instead, most theropod material from Transylvania consists of small teeth belonging to several kinds of small (2 m long), yet aggressive killers roaming the Hațeg region in the Late Cretaceous (Plate IV, top). Troodontids were also small predators, with long, slender, lightly built skulls that housed a large brain and very large eyes (positioned in such a way as to give these dinosaurs binocular vision) and was equipped with many small, but sharply recurved teeth. Alvarezsaurids, a group known principally from the Late Cretaceous of Mongolia, Argentina, and the United States, were small (about 1 m long), cursorial theropods with exceedingly diminutive, but robust forelimbs and long, gracile hind limbs.[50] The phylogenetic position of alvarezsaurids within Theropoda is presently controversial: they have been placed within or just outside Avialae or as the sister group of Ornithomimosauria. Sauropods such as *Magyarosaurus* and other herbivorous inhabitants of the Transylvanian region surely would have feared an attack by any or all of these predators. The form and lifestyles of *Euronychodon*, *Paronychodon*, and *Richardoestesia* are much more problematic. We know almost nothing about any of these theropods beyond their small size and the details of their sharp teeth. Nor can we say much about their affinities. *Euronychodon* has been thought by some paleontologists to be a primitive ornithomimosaur, and by others to be an indeterminate coelurosaur or even an indeterminate theropod;[51] *Paronychodon* may be a troodontid; and *Richardoestesia*, thus far, stands as a tetanuran of no known affinities. These are really only guesses, and there is little hope of understanding the evolutionary and ecological significance of these three small theropods without better material than we presently have.

The best known of all of these small theropods are the dromaeosaurids, fast-running, bipedal predators with a large, sharply curved claw on each hind foot. The best known of the Transylvanian dromaeosaurids is the newly discovered *Balaur bondoc* ("stocky dragon" in archaic Romanian). *B. bondoc* presently consists of a partially complete, articulated skeleton from the Sebeș Formation near Sebeș (a second specimen is known from the Densuș-Ciula Formation at Tuștea).[52] Unlike many other dromaeosaurids, this theropod is peculiar in having short fore-

Figure 2.15. Left lateral view of what is known of the skeleton of *Balaur bondoc*. Scale = 10 cm. (After Csiki et al. 2010)

limbs, highly fused hands and distal hind limbs, a very retroverted pubis, and dual claws capable of extreme hyperextension on each foot, presumably for grasping or disemboweling prey (figure 2.15). Similar in size to contemporary Laurasian dromaeosaurids, *B. bondoc* is presently the best known theropod from the Late Cretaceous of Europe. It is thought that dromaeosaurids in general, and perhaps *B. bondoc* in particular, may have hunted in packs, dispatching their prey by leaping upon them, raking them open with their deadly toe claws, and using their rigid tail to maintain balance.

The Nodosaurid Ankylosaur *Struthiosaurus transylvanicus*

Ankylosaurs are one of the two major groups of ornithischians with bony plates on their back (the other being Stegosauria), and nature lavished a full suit of armor on them.[53] Ankylosaurs were probably experts in hunkering down in self-defense.

The first-discovered ankylosaur was *Hylaeosaurus*. One of the original members of Owen's Dinosauria, it was not clear, at the time of its discovery in 1832, what this bizarre animal truly looked like.[54] Nor was it much clearer to Franz Nopcsa when the armored dinosaur—*Struthiosaurus transylvanicus*—was first discovered in the Haţeg Basin in 1912. At that time, few good ankylosaur specimens had been uncovered anywhere in the world. It took several more years and many discoveries for paleontologists to conclude that these were lumbering quadrupeds covered by a shell-like armor of bony plates and spines across the neck, back, and tail. With these discoveries, paleontologists also determined that

BOX 2.3 **Evolutionary Relationships in Ankylosauria**

Ankylosauria was first recognized as a group in 1923 by Henry Fairfield Osborn of the American Museum of Natural History in New York. These lumbering, armor-bedecked quadrupeds are now known to consist of two groups, more or less equal in diversity (14 or so species each): Nodosauridae and Ankylosauridae. Recent phylogenetic research by Matt Vickaryous and his coworkers, and by Atilla Ősi and László Makádi, has provided the basis for understanding their evolutionary history.

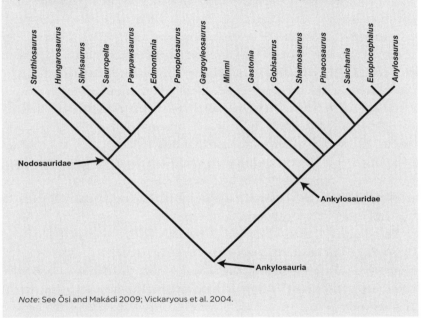

Note: See Ősi and Makádi 2009; Vickaryous et al. 2004.

ankylosaurs are principally formed into two groups: nodosaurids, with their great shoulder spines, and club-tailed ankylosaurids (box 2.3).

The story of *Struthiosaurus* is as intricate and complex as that of the Transylvanian theropods. It begins in 1871, when Emmanuel Bunzel, an Austrian physician and avid fossil collector, described new vertebrate material recovered from the Gosau Beds of Muthmannsdorf, near Vienna.[55] One specimen consisted of a small braincase fragment that he thought was rather birdlike; Bunzel named the specimen *Struthiosaurus austriacus* ("Austria's ostrich-reptile"). However, Harry Govier Seeley first identified the ankylosaurian nature of *Struthiosaurus* in 1881.[56] Thereafter, when Nopcsa studied the ankylosaur material from the Hațeg Basin, he compared it closely with *S. austriacus* and the other ankylosaurs from Europe. On the basis of these studies, Nopcsa considered the Hațeg ankylosaur to be generally the same as Bunzel's *Struthiosaurus*, yet sufficiently different to merit a new species designation, *Struthiosaurus transylvanicus*.[57]

S. transylvanicus is currently known from skull elements (the braincase and portions of the skull roof and cheek region), as well as vertebrae, the shoulder girdle, and dermal armor belonging to at least two individuals collected from the Sânpetru Formation in the Hațeg Basin (figure 2.16). There is new material of *S. transylvanicus*, collected at Oarda de Jos and Vurpăr by researchers from Universitatea Babeș-Bolyai Cluj Napoca, which should provide additional information on the anatomy of this nodosaurid.[58] All of these specimens are relatively small, but they are thought to be from adult individuals, suggesting that members of this species grew to not much more than 2 m in length. What we know about *Struthiosaurus* is due in large part to studies in the 1990s of *S. austriacus* by Xabier Pereda-Suberbiola, a Basque dinosaur paleontologist also working in Paris, and Peter M. Galton, a paleontologist and anatomist at the University of Bridgeport in Connecticut.[59]

The skull of *S. transylvanicus* is robustly built: most of the holes at the back of the head, through which the jaw muscles could bulge, are reduced or closed, the back surface of the skull is fused into a single unit, and the top of the head is covered with armor. Although their placement and appearance are yet unknown, armor plates also covered the back of the body, and a large spine projected from its shoulder. In addition, the outer surface of the shoulder girdle bears a prominent ridge (pseudoacromion), where strong muscles were attached to move the large and muscular forelimb.

Paleontologists have relied on these and other features of *Struthiosaurus* to provide clues to its kinship and lifestyle (Plate IV, bottom). It is clearly a nodosaurid, based on the scapula having a pseudoacromion that is displaced downward and backward toward the shoulder joint and ends in a knoblike projection.[60] Nopcsa thought that there may have been teeth in the front of the upper jaws (known as premaxillary teeth), unlike more derived nodosaurids such as *Edmontonia* and *Panoplosaurus*,

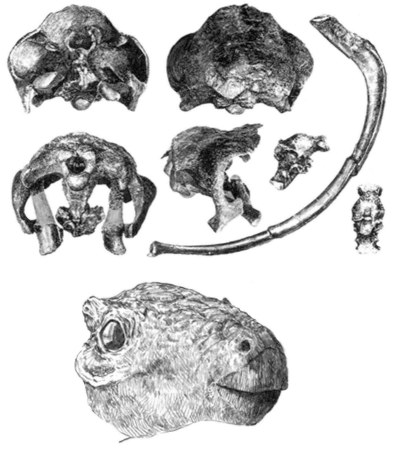

Figure 2.16. The Transylvanian nodosaurid ankylosaur, *Struthiosaurus transylvanicus* (*above*), and a reconstruction of its head (*below*). (*Struthiosaurus* from an original plate from Nopcsa 1929a; restored head after *Pawpawsaurus* and *Edmontonia*)

whose premaxillary teeth are absent. If this is true, *Struthiosaurus* would assume a relatively primitive position in nodosaurid phylogeny.

Given its size and posture, we assume *Struthiosaurus* was a very low-browsing feeder, foraging no more than a meter above the ground. The relatively narrow and scoop-shaped front of the upper and lower jaws of other nodosaurids (but not yet known in *Struthiosaurus*) was probably overlaid with a horny covering (called a rhamphotheca), also seen in living turtles and birds, and suspected in many ornithischian dinosaurs. The shape of this region of the head suggests that these animals were somewhat selective feeders, plucking or biting at particular kinds of leaves and fruits with the sharp edge of the rhamphotheca.[61]

Food, once cropped, was apparently chewed through a combination of simple up-and-down puncturing and fore-and-aft grinding. Beyond this suggestion, which we deduce from the pattern of wear on the teeth, ankylosaur mastication is puzzling. For example, the triangular teeth of both nodosaurids and ankylosaurids do not appear particularly well suited to a diet of plants; they are small, not very elaborate, and less tightly packed together in the jaws than the teeth of other ornithischian dinosaurs. On the basis of these simple teeth, Nopcsa thought that *Struthiosaurus* instead fed on insects,[62] although no one today doubts that its diet was dominated by plants. An extensive bony palate closed off the oral cavity from the nasal cavity, allowing these ankylosaurs to chew and breathe at the same time. Moreover, deeply inset tooth rows suggest the presence of deep cheek pouches to keep food from falling out of the mouth. The jaw bones themselves were relatively large and strong, although lacking enlarged areas for muscle attachment. Every jaw feature except the teeth suggests that ankylosaurs were adept chewers.

Perhaps the paradox of unsophisticated teeth set in strong, cheek-bound jaws can be understood by looking not at how ankylosaurs chewed food before swallowing, but at the other end of the animal, where much of the plant digestion must have been accomplished by gut fermentation. What the teeth couldn't accomplish mechanically, the gut could, by breaking down the roughly chopped leaves by chemical means. A deep, broadly rounded rib cage circumscribing an enormously expanded abdominal region indicates that the digestive tract was huge.

Nopcsa's interest in *Struthiosaurus* extended beyond how and what it ate to how it moved and how its small brain controlled the movement of its great mass. We know the brain size of this Transylvanian nodosaurid from a latex cast of the brain cavity that Nopcsa had made for his detailed study of the beast. Measuring less than 50 ml, the brain of *Struthiosaurus* was very small, even by dinosaur standards—only sauropods had smaller brains for their body size. There were no intellectual giants here, nor was athleticism their strong point. In addition, *Struthiosaurus*, like other ankylosaurs, were among the slowest moving of all dinosaurs. According to Australian paleontologist Tony Thulborn's calculations,[63] these armored dinosaurs walked at a leisurely pace of about 3 km/hour and probably ran no faster than 10 km/hour, about the average running speed of an elephant among living animals. *Struthiosaurus* was plodding, to be sure, but certainly not without defenses. It probably was able to stab at predators and competitors alike by firmly planting its hind limbs, ducking its head, and rolling its strong shoulders, with their formidably long spines, forward. Otherwise, *Struthiosaurus* hunkered down, relying on a suit of bony armor to protect it from packs of troodontid or dromaeosaurid theropods.

The Transylvanian Ornithopods: *Zalmoxes*

Zalmoxes, named for the Dacian god of the underworld and immortality who was famous in Romanian lore for living, teaching, and healing from his subterranean crypt, was one of two sorts of Transylvanian ornithopods. Our account of *Zalmoxes* and how it fits into the ornithopod scheme of things begins in 1897, when Nopcsa first recognized some distinctive bones and teeth recovered from localities along the banks of the Sibișel River.[64] These he called *Mochlodon*, because of their similarity to ornithopod specimens from the Gosau Beds of Austria (the same locality that yielded the bones of *Struthiosaurus austriacus*) that Seeley had named *Mochlodon* in 1883.[65] *Rhabdodon priscus*, another Late Cretaceous dinosaur from southern France originally described in 1869 by Philippe Matheron, a geologist from Marseille,[66] was also similar to the Hațeg *Mochlodon* material, but Nopcsa initially regarded them as different kinds of ornithopods. By 1915, however, he was considering the possibility that the two were from the same species, perhaps being male and female. Since *Rhabdodon* was the earlier find, Nopcsa conceded Matheron's claim of priority and renamed the *Mochlodon* material from Transylvania *Rhabdodon*, for Matheron's animal.[67]

Once *Rhabdodon* was properly named to Nopcsa's satisfaction, where did it belong on the ornithopod family tree? Nopcsa was the first to advocate particular affinities for this Hațeg ornithopod in 1902. Comparing it with other ornithopods, he noted that *Rhabdodon* seemed to be more primitive than *Iguanodon* (from the Early Cretaceous of Europe), but more closely related to *Camptosaurus* (from the Late Jurassic of North America). Thereafter, *Rhabdodon* was either included in a group that contained *Iguanodon* (either Iguanodontidae or the more encom-

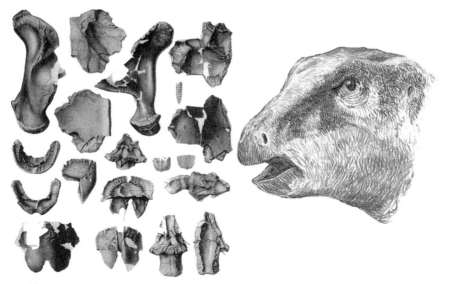

Figure 2.17. Skull elements of *Zalmoxes*, one of the Transylvanian ornithopods (*left*) and a reconstruction of its head (*right*). (Skull elements from an original plate from Nopcsa 1904)

passing Iguanodontia) or shifted to the group of small ornithopods called Hypsilophodontidae.[68]

Although Nopcsa regarded the proper parentage of *Rhabdodon* as solved, from the perspectives of both taxonomy and evolutionary relationship, we were far from convinced when we started examining the remains of this beast. So, working closely with Zoltán Csiki (Universitatea din București) and David Norman (Sedgwick Museum at Cambridge University), we tried to bring the animal Nopcsa ended up calling *Rhabdodon* (figure 2.17) into its present-day context. We combed through dusty museum drawers across Europe, trying to compare our Romanian specimens with material of *Rhabdodon* from France and Spain, and *Mochlodon* from Austria, as well as with other specimens from England and North America (*Camptosaurus*, *Hypsilophodon*, and *Tenontosaurus*, among others), integrating this trove of data using cladistic analyses (box 2.4).

Here's what we discovered during the course of our research. The specimens from France and Spain that have been referred to as *Rhabdodon* probably all belong to the same genus and species, a conclusion

BOX 2.4 Evolutionary Relationships in Ornithopoda

A little more than 160 million years ago, the Middle Jurassic saw the emergence of relatively small (2 m long), bipedal, herbivorous dinosaurs called ornithopods. From the Late Jurassic, and especially through the Cretaceous, these ornithopods were among the dominant terrestrial forms around the world. Some ornithopods, such as *Hypsilophodon* and *Orodromeus*, were small (approximately 2 m long) and fast running, while others, like the larger iguanodontians, are probably the most familiar of all ornithopods. Their namesake, *Iguanodon*, was one of the first-discovered dinosaurs and is a charter member of Owen's Dinosauria. This 10 m long ornithopod from the Early Cretaceous of Europe had a long, straight tail, a stocky body, and large spikes serving as thumbs. Hadrosaurids, often called duck-billed dinosaurs, are among the remaining iguanodontoideans; they look like an unlikely evolutionary cross between a duck and a horse. They, too, probably had muscular cheeks, but with strong and closely packed upper and lower teeth that formed roughened grinding surfaces.

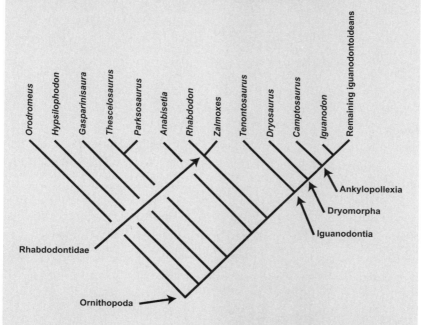

Note: See Horner et al. 2004; Norman 2004; Norman et al. 2004; Weishampel et al. 2003. See especially Butler et al. 2008 for the most recent assessment of the taxonomic composition of Ornithopoda.

that was confirmed by material and analyses by Marie Pincemaille and her colleagues at Université Montpellier in the late 1990s (figure 2.18).[69] This creature naturally retains the name proposed in 1869 by Matheron, *Rhabdodon priscus*. However, our work also identified two species from Transylvania, both of which are now known from various localities in the Hațeg Basin and several other locales—Vurpăr, Oarda de Jos, Jibou, Bărăbanț, Lancrăm, Sebeș—in the Transylvanian Depression.[70] In fact, the two Transylvanian species appear more closely related to each other than either is to *Rhabdodon* from France and Spain. We considered this a prime reason for coining a new name, and in 2003 we settled on *Zalmoxes*. The two species were called *Zalmoxes robustus* and *Zalmoxes shqiperorum* ("shqiperorum" honoring Nopcsa's love of the northern Albanian tribes).[71] Of the two, *Z. robustus* is quite well known, based on upwards of 300 specimens, whereas *Z. shqiperorum* is known from a nearly complete skeleton from Nălaț-Vad, as well as a few dozen bones and teeth from elsewhere in the Hațeg Basin. It is further known from the Șard Formation at Vințu de Jos on the southern margin of the Transylvanian Depression,[72] and from the Jibou Formation near Jibou along its northern margin.[73]

So *Zalmoxes* has now received its name, but how is it related, phylogenetically, to other ornithopods? Our cladistic analyses clearly indicated that its closest relative was *Rhabdodon*. Together, *Zalmoxes* and *Rhabdodon* form a clade called Rhabdodontidae, which is most closely related to the group of ornithopods known as iguanodontians, the latter including animals such as *Tenontosaurus*, *Iguanodon*, hadrosaurids, and many other forms.

Now that we can recognize *Zalmoxes* as distinct from *Rhabdodon* and other ornithopods, we can also begin to understand its form and lifestyle (Plate V, top). The bones suggest a moderate-sized ornithopod (3–4 m long). Its body was stocky, it had a bipedal stance, and its back was held nearly horizontally, with its long muscular tail counterbalancing the front half of the body at the hip joint. The features of the bones of the fore- and hind limbs suggest that these regions were well muscled, whereas the ribcage seems more barrel-shaped than in other ornithopods. Limb proportions indicate that *Zalmoxes* was probably not a particularly fast runner, scurrying along at no more than 25 km/hour.[74] Although it would be great if we had numerous trackways of *Zalmoxes* to

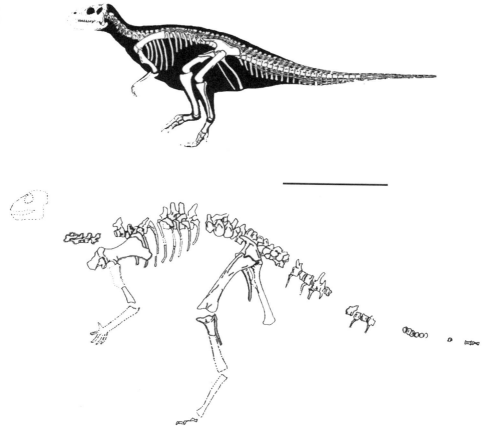

Figure 2.18. Skeletal reconstructions of *Zalmoxes* (*above*) and *Rhabdodon* (*below*). Scale = 100 cm. (*Rhabdodon* after Garcia et al. 1999)

test these calculations, there is a track site in the Vurpăr Formation near Sebeș (the only footprints known from the Late Cretaceous of Transylvania) that consists of two hind footprints thought to have been made by *Zalmoxes*. These tracks indicate a more leisurely walking speed of less than 7 km/hour.[75]

The skull of *Zalmoxes* is relatively large but compact, with a short face (figure 2.19). Some individuals have a stout transverse crest above the eyes, perhaps a sign of sexual dimorphism or a sign of age. The narrow, toothless beak was probably covered in life with a sharp rhamphotheca (as also discussed for *Struthiosaurus*), and was thus likely to have been

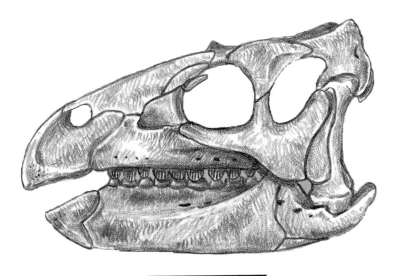

Figure 2.19. A reconstructed skull of *Zalmoxes robustus*. Scale = 10 cm. (After Weishampel et al. 2003)

able to cut through tough foliage and fruits. Both the upper and lower jaws were strongly built, containing at least ten large, closely packed teeth. Patterns of tooth wear reveal that *Zalmoxes* chewed its food well, using a transverse grinding motion. The teeth are recessed from the sides of the face, suggesting—as in the case of *Struthiosaurus* and other ornithischians—that cheeks covering this region may have prevented food from slipping out of the sides of the mouth.

We don't know much about the sociality, growth and development, and reproductive behavior of *Zalmoxes*. On these matters, the fossil record has been silent. We do know that this ornithopod is probably the most common element within the Transylvanian assemblages, but whether it was truly gregarious or merely appeared frequently amid the fauna is unknown. Similarly, there are no data to suggest colonial nesting and parental care—or rule it out. For these aspects of a dinosaur's life, we need to turn to the Transylvanian hadrosaurids.

The Transylvanian Ornithopods:
Telmatosaurus transsylvanicus

The duckbills of the Late Cretaceous—members of Hadrosauridae—were among the most diverse forms of plant-eating dinosaurs, known principally from North America and Asia (figure 2.20). Many hadrosaurids are known from well-preserved fossils, including those of embryos, hatchlings, juveniles, teenagers, and adults. From these specimens, and from abundant nesting sites, we are learning more about parental care of offspring, group nesting, and the rapid growth rates that appear to characterize these ornithopods.

As we have noted previously, the first dinosaur described from the Hațeg Basin was a new genus and species of hadrosaurid dinosaur that Nopcsa named *Telmatosaurus transsylvanicus*. Virtually the entire skull and the majority of the postcranial skeleton of *T. transsylvanicus*, based on more than 100 cranial and postcranial elements, are known from a number of individuals of various body sizes collected at numerous localities in the Hațeg Basin and Transylvanian Depression.[76] At an adult length of about 5 m, *Telmatosaurus* was one of the smallest hadrosaurid dinosaurs (Plate V, bottom), much smaller than hadrosaurids elsewhere in the world or their more distant relatives (such as *Iguanodon*), both of which ranged upwards of 10 m in length.[77] We will return to this issue of body size and its evolutionary significance in chapter 6.

Even though the original material of *Telmatosaurus* that Nopcsa described in 1899 was severely crushed, we were able to compare it with other, subsequently discovered specimens and thereby reconstruct what the skull must have looked like in three dimensions (figure 2.21). What emerged from the reconstruction is a long, somewhat horselike cranium, reminiscent of *Iguanodon* and other hadrosaurids.[78] The front of the snout (the premaxilla) is narrow, toothless, and crenulated, most probably supporting a rhamphotheca. Both the upper and lower jaws contain as many as 30 vertical positions for the teeth, far more toothy than the jaws of non-hadrosaurid iguanodontians, but considerably less so than in most other hadrosaurids. Both the upper and lower sets of teeth consisted of hundreds of functional and replacement teeth, interlinked to form dental batteries (figure 2.22). Such a complex arrangement of teeth, coupled with well-developed jaw muscles and a unique mastica-

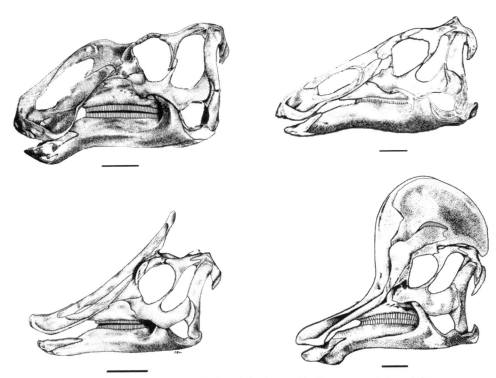

Figure 2.20. Diversity of skull morphology in hadrosaurids: *Gryposaurus* (above left), *Edmontosaurus* (above right), *Saurolophus* (below left), and *Corythosaurus* (below right). Scale = 10 cm. (After Weishampel and Horner 1990)

tory system, would have made short work of the toughest vegetation.[79] Behind the head, the skeleton of *Telmatosaurus* looks much like that of both more primitive iguanodontians (such as *Ouranosaurus* and *Iguanodon*) and other hadrosaurids (like *Brachylophosaurus* and *Maiasaura*).[80] However, its body was smaller than these taxa, and therefore not unduly bulky. Similar to other ornithopods (including *Zalmoxes*), the long tail, stiffened and strengthened by crisscrossing bony tendons, acted to counterbalance the front half of the body.

In 1993, *Telmatosaurus* was subjected to cladistic analyses to see where it fit within hadrosaurid evolution (box 2.5).[81] Based on this work, which has been borne out in subsequent studies,[82] *Telmatosaurus* appears to fit best as the sister group of both lambeosaurines and hadrosaurines. In 1993, Weishampel et al. called this latter clade Euhadro-

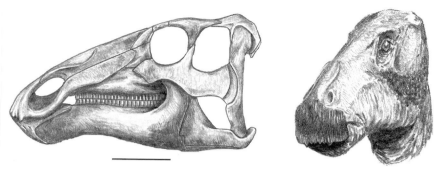

Figure 2.21. A reconstruction of the skull of *Telmatosaurus transsylvanicus* (*left*) and a reconstruction of its head (*right*). Scale = 10 cm

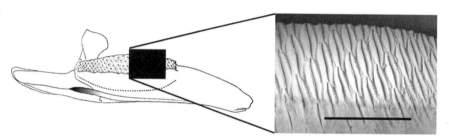

Figure 2.22. A hadrosaurid dental battery (*left*), indicating the complex interlocking pattern of replacement teeth (*right*). Scale = 5 cm

sauria, but most paleontologists have followed a more historical path, considering hadrosaurines and lambeosaurines as the sole members of Hadrosauridae; this makes *Telmatosaurus* a non-hadrosaurid hadrosauroid. In either taxonomy, the relationships remain the same: *Telmatosaurus* is a (or perhaps the most) primitive outsider to all other hadrosaurids. We will discuss the significance of the position of *Telmatosaurus* with respect to ornithopod evolution, especially in relationship to its small body size, later in this book.

Large ornithopods in general were not particularly fleet of foot, and *Telmatosaurus* was no exception. Based both on hadrosaurid trackways and on limb proportions, these bipedal forms probably were able to reach a top speed of 15–20 km/hour during a sustained sprint, but at

BOX 2.5 Evolutionary Relationships in Hadrosauridae

Hadrosaurids, arguably the most important group of herbivorous dinosaurs at the end of the Cretaceous, especially in North America and Asia, are now known from more than 30 species. Traditionally, these have been placed in two subgroupings: hadrosaurines, with a solid crest or a flat head, on the one hand, and hollow-crested lambeosaurines on the other. Cladistic analyses of these species demonstrate the monophyly of both Lambeosaurinae and Hadrosaurinae (although support for the latter is not strong), which together form the larger Euhadrosauria. In addition, an ever-increasing number of taxa do not fit into the euhadrosaurian clade. Among them is the Transylvanian hadrosaurid *Telmatosaurus transsylvanicus*. Another is *Bactrosaurus johnsoni*, from the Late Cretaceous of northern China and southern Mongolia. *Bactrosaurus* has been interpreted as a lambeosaurine hadrosaurid and as a non-hadrosaurid hadrosauroid. Here we will treat *Bactrosaurus* as an unresolved basal hadrosaurid.

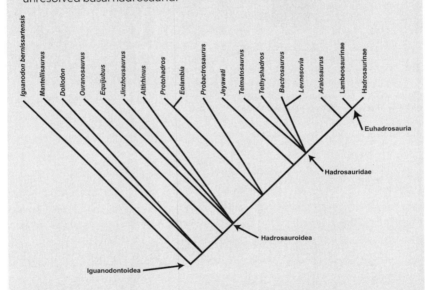

Note: For Euhadrosauria, see Horner et al. 2004; Weishampel et al. 1993. For non-hadrosaurid hadrosauroids, see Godefroit et al. 1998.

slower speeds and at rest, they apparently assumed a quadrupedal posture. In this position, *Telmatosaurus* most likely browsed no higher than a meter or so above the ground, probably using its forelimbs to grasp at leaves and branches in order to bring the foliage closer to its mouth.

From *Gryposaurus* (with its arched snout), to *Saurolophus* (with its supracranial spine), to *Parasaurolophus* (with its hollow, plumelike crest), these dinosaurs stand out from the crowd with their wild headgear. In 1975, Jim Hopson of the University of Chicago examined the functional significance of these cranial adornments. He argued that the arches, spikes, hollow crests, and other cranial decorations evolved in the context of complex intraspecific social behavior. These crests were used in intraspecific aggression and display, both visual and vocal; in courtship displays; and in mate rivalries. Crests and the rest presumably helped hadrosaurids recognize kin, avoid enemies, display to each other and to members of different species, communicate with offspring, and establish social hierarchies—an amazing evolutionary achievement that placed hadrosaurids in the top ranks of complex social behavior among dinosaurs.

As a basal member of the group, *Telmatosaurus* may have something to tell us about the origin of these aspects of hadrosaurid sociality. New *Telmatosaurus* material indicates that this dinosaur had a pair of sinuous ridges that ran along the sides of the snout. Though less dramatic than the headdresses of more evolved hadrosaurids, these ridges would have made its face more visually interesting than those of more primitive ornithopods. Were the ridges parts of a display apparatus, perhaps a progenitor to the exhibitionists yet to come? Were these sigmoidal ridges sexually dimorphic? We don't yet know. Nor can we say with certainty whether *Telmatosaurus* was territorial, or engaged in parental behavior. The only other aspect of behavior known in this Transylvanian hadrosaurid is an ever-so-brief glimpse into its early growth and development, which comes from the discovery of embryo or baby bones associated with some eggs unearthed in the late 1980s.

The Haţeg Dinosaur Egg Nests

Nests and nesting horizons are among the most exciting dinosaur discoveries in Romania over the past decade and a half. The first announcement of Haţeg Basin eggs came in 1989, through work near the village of

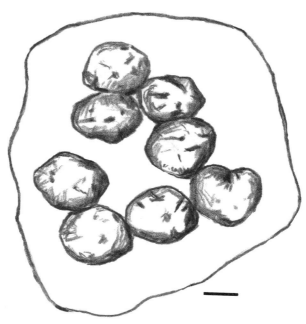

Figure 2.23. A reconstruction of the dinosaur nest from Tuștea. Scale = 15 cm. (After Grigorescu et al. 2010)

Tuștea by researchers from Universitatea din București.[83] Another site, Totești-baraj, was discovered by the joint expedition from Universitatea Babeș-Bolyai Cluj Napoca, Romania, and Brussels' Institut Royal des Sciences Naturelle de Belgique in 2000.[84] The eggs within these nests seem to be laid in a somewhat organized, curvilinear fashion.[85] In life, the eggs would have been nearly spherical, approximately 15 cm in diameter, and nearly a liter in volume (figure 2.23). On average 2.4 mm thick, the eggshell is covered with an irregular pattern of small, hemispheric tubercles. This locality has thus far produced forty eggs distributed in eleven different nests. It is not certain what the conformation of each nest is, but the eggs themselves, although deformed, also appear to be subspherical in shape, with a diameter of about 15 cm.

In addition to their macroscopic features, still more information can

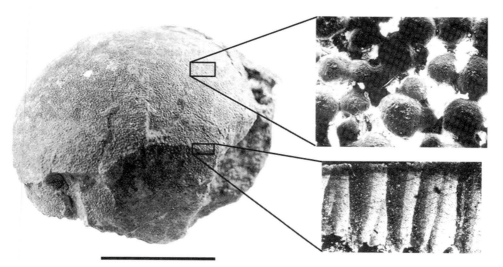

Figure 2.24. One of the eggs recovered from Tuștea (*left*), indicating the surface texture (*above right*) and a shell cross-section (*below right*). Scale = 10 cm. (After Weishampel et al. 1991)

be discerned about these eggs under the microscope (figure 2.24). At a magnification of about 80 power, the eggshell reveals its mineral structure. The eggshell is calcite, organized into structural units that vary characteristically among different egg-laying species. Circa the early 1990s, Karl Hirsch and Konstantin Mikhailov erected a classification scheme for these structural units.[86] They identified basic types, each broken down into subgroupings to produce a taxonomy of names that trip up the most dexterous tongue: dendrospherulitic, angusticanaliculate, Laevidoolithidae. Still, it's often difficult to attribute these categories to particular kinds of egg-laying animals, in large part because we generally lack a direct connection between eggs and moms. However, some truly remarkable occurrences—embryonic remains preserved in fossil eggs—have proven to be a Rosetta Stone. Consequently, we know that eggs found in Montana, which fall within the Mikhailov/Hirsch category termed Spheroolithidae, were laid by the hadrosaurid called *Maiasaura*, because they were found with the embryonic or hatchling bones of this dinosaur.[87] On a similar basis, we know that oviraptorid theropods laid elongatolithid eggs, and, based on discoveries in Argentina made in 1998, titanosaur sauropods laid megaloolithid eggs.[88]

Based on their architecture, the Tuştea eggs likewise belong to the Megaloolithidae category, sharing similarities in microstructure, pore organization, size, and shape with some of the eggs from southern France. The only principal difference is in nest structure: the French megaloolithid eggs apparently were laid in sweeping curves instead of curvilinearly.[89] Although no embryonic remains associated with these eggs have ever been recovered in France, they are thought to have been laid by a titanosaur sauropod, an animal of the right size to have laid such an egg and one common in the Late Cretaceous fauna of the region. Realizing that we also had sauropods in the Hațeg fauna, at first we suggested, in 1990, that these eggs may have been laid by one or the other titanosaur from Transylvania.[90]

Even more significantly, the bones of either full-term embryos or newborn hatchlings (technically called perinatal remains) were also discovered at the same site, in fact from the same bedding planes that yielded the Tuştea egg clutches. Consisting of partially articulated skeletons and additional associated remains, the specimens can be identified with confidence as those of a hadrosaurid (most probably *Telmatosaurus*), even though the joint surfaces of the bones are very porous and the surface texture of their shafts is immature (figure 2.25).[91] Their proximity to the Tuştea eggs suggested that the perinatal remains belong with the eggs; that is, the female who laid the eggs would have been the mother of the babies whose small bones we found. This question of who laid the Tuştea clutches has brought some consternation to dinosaur "egg-ologists": the structure of the Tuştea eggs resembles that of titanosaur sauropods (thereby implicating *Magyarosaurus* as the egg-layer), whereas the perinatal fossils come from a hadrosaurid (suggesting that the parent was *Telmatosaurus*).[92] This conflict might be solved if we found identifiable embryonic remains preserved *inside* the Tuştea eggs and not alongside them. Some scanning electron microscopic photographs have been taken,[93] but they are far from conclusive.

This conundrum—titanosaur or hadrosaurid eggs—has been indirectly solved by discoveries made halfway around the world, in the badlands of Patagonia in the southern half of Argentina.[94] It was here, at a site known as Auca Mahuevo, that an extensive nesting ground—covering more than a square kilometer and littered with tens of thousands of large, unhatched eggs—was discovered in 1997 by an expedition led by Luis

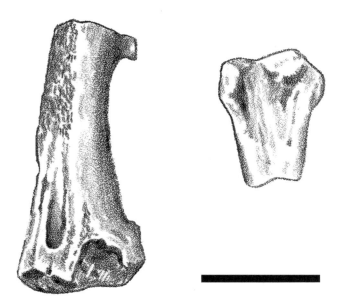

Figure 2.25. Perinatal hadrosaurid remains from Tuştea: a distal femur (*left*) and a proximal tibia (*right*). Scale = 10 mm. (After Weishampel et al. 1993)

Chiappe from the Natural History Museum of Los Angeles County, Lowell Dingus from the American Museum of Natural History in New York, and Rodolfo Coria from the Museo Municipal Carmen Funes in Plaza Huincal, Argentina. Unlike the Haţeg clutches, those from Argentina were organized into clusters of between 15 and 34 eggs. Most spectacularly, and of particular relevance here, was the fact that a high proportion of these eggs contained embryonic skeletons, some with the impressions of embryonic skin! Moreover, these imprisoned embryos, many of them in near articulation, possess nearly complete skulls that clearly belong to titanosaur sauropods.

The importance of Auca Mahuevo to our Transylvanian research is at least threefold. First, we have an unambiguous association of eggs with embryos. Indeed, they are the first known embryonic remains of sauropods. Thus the question of who laid these clutches has been answered with absolute certainty: titanosaurs. Second, the incredible geographic extent of the nesting horizons at Auca Mahuevo speaks strongly for colonial nesting and maybe for gregarious behavior in these and perhaps other titanosaurs elsewhere in the world. Finally, these titanosaur embryos provide the opportunity to discover the details of early stages of sauropod growth and development, a subject we will take up in chapter 7.

Returning to Hațeg, what we have learned from Auca Mahuevo is that the clutches of these Transylvanian eggs were most probably laid by one or the other titanosaur from the fauna. From the sediments surrounding the nests, we known that the eggs were buried under a thin layer of fine sand. From the construction of the eggshell, we can calculate the rate of water–vapor exchange, absolutely critical during embryonic incubation, and we can conclude that, for optimal conditions, the humidity of the nest should have been between 85% and 95%, and the eggs would have hatched in 50–60 days.[95] Not a bad picture of nesting paleoecology with only data from pores, shell thickness, and sedimentology to go by.

The perinatal bones known from Hațeg, on the other hand, may have washed into the titanosaur nesting area, but from very close by, because their porousness and delicacy would have made long-distance transport impossible. These small bones belong to a hadrosaurid, probably *Telmatosaurus transsylvanicus*, and reveal different aspects of how this ornithopod may have grown and developed. Because hadrosaurids (including *Telmatosaurus*) are skeletally immature when compared with other perinatal dinosaurs, it appears that they hatched at an underdeveloped, dependent stage. Jack Horner, the Museum of the Rockies (Bozeman, Montana) paleontologist whose work has revolutionized our understanding of dinosaur growth and development, has argued that these immature hatchlings would have been nest-bound while they matured.[96] This extended parental care, necessary to promote their survival, is a condition ecologists call altriciality.[97] *Telmatosaurus* was almost certainly an altricial dinosaur, and, given its primitive position among hadrosaurids, it could represent the turning point in the origin of this hadrosaurid life-history strategy. We will have more to say about this connection in chapter 7.

CHAPTER 3

Pterosaurs, Crocs, and Mammals, Oh My

The array of Late Cretaceous creatures from Transylvania goes well beyond its dinosaurs. Although they were the most obvious animals in the Hațeg landscape, *Zalmoxes*, *Struthiosaurus*, *Magyarosaurus*, *Telmatosaurus*, and others were among the rarest members of the fauna. Just as every ecosystem has its myriad players—primary producers, herbivores, predators, decomposers, and many more, so we must look for more than just the dinosaurs of Transylvania for a more complete picture of life in this region some 72 million years ago. Who were these other creatures also making a living there?

To answer this question, we turn to all of the non-dinosaurian members of the Transylvanian vertebrate assemblage known to date. To cover them in a reasonably logical sequence, we order our discussion around their kinship to dinosaurs (figure 3.1). As pterosaurs are the amniotes (land-dwelling vertebrates) most closely related to dinosaurs, we begin with these flying reptiles. Thereafter, we turn to crocodilians, squamates (lizards and snakes), turtles, and mammals. Finally, we will step outside of amniotes to cover the amphibians and fish from the Late Cretaceous of Transylvania.

THE HAȚEG PTEROSAURS

Pterosaurs have a unique position in the Mesozoic world—they were the first fully flying tetrapods, establishing themselves as masters of the air

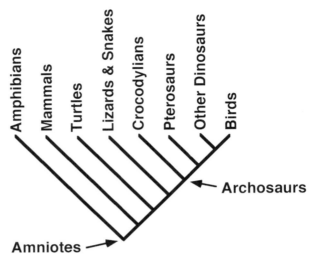

Figure 3.1. Evolutionary relationships of Amniota

as early as the Late Triassic and dominating the skies throughout the world for over 170 million years. Much of what we know about these flying reptiles comes from the Jurassic of Europe, but the Cretaceous pterosaur record indicates their presence on all continents, with the exception of Antarctica.[1] By the Cretaceous, a variety of long-tailed forms had come and gone, to be replaced by short-tailed Pterodactyloidea, which were to become the largest flying animals—some species with a wingspan of nearly 12 m—ever to evolve.

Wonderfully preserved bones of these flying behemoths have been recovered in abundance from Cretaceous beds in North and South America, although specimens from the Late Cretaceous of Europe are quite rare, known from only three localities: Choceň in northeastern Czechia (Czech Republic), Muthmannsdorf in eastern Austria, and Sânpetru and Vălioara in Romania. In 1881, Anton Fritsch, a Czech paleontologist, described some new fossils that he called *Cretornis*. Although he originally thought that they belonged to a bird, some small, well-preserved wing bones were collected that helped him identify this animal as a pterosaur. Fritsch then referred his *Cretornis* material to *Ornithocheirus* as a new species, *O. hlavatschi*, named after the pharmacist Hlaváč in the town of Choceň who had collected the fossils.[2] Two years later, Seeley recognized the first pterosaur material (an articular bone from a lower jaw that

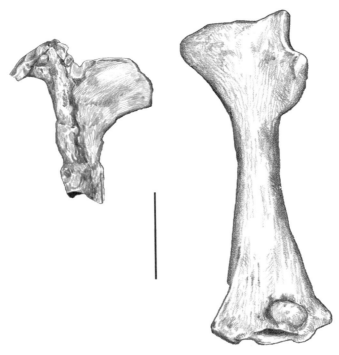

Figure 3.2. Humeri of *Ornithocheirus bunzeli* (*left*) and *Ornithocheirus hlavatschi* (*right*). Scale = 2 cm

Bunzel had interpreted as belonging to a lizard) from the Gosau fauna of Muthmannsdorf, Austria, calling it *Ornithocheirus bunzeli* (figure 3.2).[3]

Even though all of these forms were called *Ornithocheirus*, the first known pterosaur bearing this name didn't come from Austria, Czechia, or Romania. *Ornithocheirus* itself was originally described by Seeley on the basis of fossils found in mid-Cretaceous rocks in England.[4] No complete skeletons of this pterosaur have ever been found, but we know from fragments of the skull, individual limb bones, and vertebrae that *Ornithocheirus* was a large pterosaur with a long, slender skull and probably a bony crest on the snout (figure 3.3). The jaws were lined from front to back with sharp teeth, suggesting that it was probably a fish eater.

Figure 3.3. A reconstruction of *Ornithocheirus*

Nopcsa announced the first pterosaur from the Hațeg fauna in 1914.[5] This specimen consists of several fused vertebrae in the shoulder region, which in some pterosaurs and birds is known as a notarium. In 1926, Nopcsa again mentioned this pterosaur in a discussion of other Late Cretaceous pterosaur material from Europe, but this time he provided further taxonomic information.[6] He regarded the Hațeg pterosaur as a member of Ornithocheiridae, perhaps related to the poorly known *Ornithodesmus* from the Early Cretaceous of England (which is now thought to be a maniraptoran theropod dinosaur).[7] Nopcsa never elaborated on these two comments, and he never described or illustrated the material. To make matters worse, the specimen was thought to be lost from the 1930s onward—that is, until 1995, when the two of us rediscovered the misplaced specimen in the collections of the Magyar Állami Földtani Intézet in Budapest.[8] Our preliminary study of this specimen indicated that Nopcsa was correct and that this was a pterosaur notarium. However, more recent work suggests that this specimen may instead be a partial sacrum of a maniraptoran theropod.[9] Whether a pterosaur notarium or a maniraptoran sacrum (it is difficult to say, and this certainly needs more study), we are fortunate in having additional

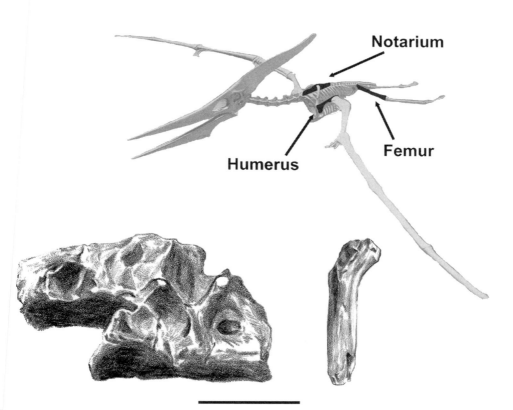

Figure 3.4. A pterosaur notarium (*bottom left*) and femur (*bottom right*) from Sânpetru. The figure of *Pteranodon* (*above*) indicates the position of these two elements and the humeri from figure 3.2. Scale = 1 cm

pterosaur specimens—a right humerus, a left femur, and possibly another notarium—more recently discovered in the same area of Transylvania (figure 3.4). On the basis of all of the Sânpetru material, we suggested that this pterosaur may have been a member of Dsungaripteroidea, but one for which we could make no further resolution (box 3.1). Unlike its closest relatives, many of which had wingspans of up to 7 m, the Hațeg dsungaripteroid must have been one of the smallest, measuring no more than 75 cm from wingtip to wingtip, about the size of many present-day bats.

The most recently discovered pterosaur material from the Hațeg Basin, even more dramatic than that previously found in Romania, was recovered from the Densuș-Ciula Formation near the village of Vălioara.

BOX 3.1 **Evolutionary Relationships in Pterosauria**

Ranging in size from a sparrow to a small airplane, pterosaurs dominated the skies of the Jurassic and the Cretaceous, becoming extinct at or near the Cretaceous–Tertiary boundary. These flyers have been placed into two groups: long-tailed rhamphorhynchoids and short-tailed pterodactyloids. However, only the latter is monophyletic, while the paraphyletic former consists of basal taxa sequentially less closely related to Pterodactyloidea.

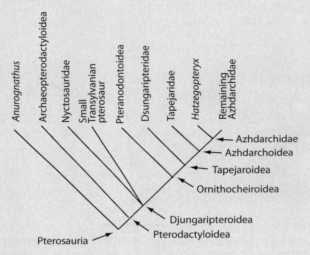

While this much is known about the phylogeny of Pterosauria, the more fine-grained relationships have been harder to come by. Recent cladistic studies by Alex Kellner (Museu Nacional in Rio de Janeiro, Brazil) have gone a long way to improve this situation. According to Kellner, pterodactyloids are divided into Archaeopterydactyloidea and Dsungaripteroidea. The former includes *Pterodactylus*, *Germanodactylus*, Ctenodactylidae, and Gallodactylidae, while the latter consists of Nyctosauridae and Ornithocheiroidea. This latter taxon includes Pteranodontoidea (*Pteranodon*, *Ornithocheirus*, and *Anhanguera*, among others) and Tapejaroidea (*Dsungaripterus*, *Noripterus*, *Tapejara*, *Quetzalcoatlus*, *Azhdarcho*, and others).

Note: See Kellner 2003.

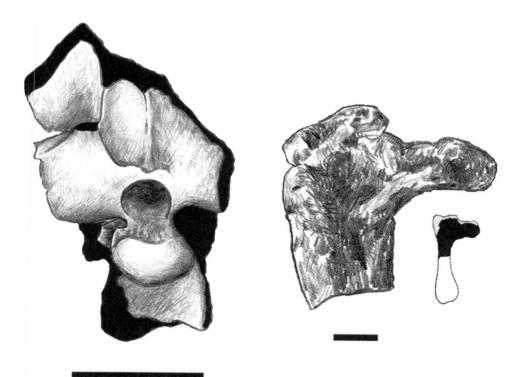

Figure 3.5. A rear view of the braincase (*left*) and a ventral view of the proximal humerus (*right*) of *Hatzegopteryx thambema*. Scale = 10 cm

Consisting only of the back end of the skull and the left humerus, these bones were originally thought to belong to a moderately large theropod.[10] However, a new study by Eric Buffetaut, Dan Grigorescu, and Zoltán Csiki has shown it to be a gigantic new flying reptile (figure 3.5), which they named *Hatzegopteryx thambema* ("the monstrous wing from Haţeg").[11] *Hatzegopteryx*, in contrast to the earlier-mentioned pterosaur, is currently thought to be a huge azhdarchid pterosaur, with an estimated wingspan greater than 12 m and a skull perhaps 3 m long, making it, if not the largest, then at least one of the largest flying animals ever to have evolved (figure 3.6).

Instead of having a wing like a bird or a bat, the wing membrane in all pterosaurs was supported solely by an elongate fourth finger. This membrane was reinforced by closely packed, parallel, flexible rods to maintain the airfoil dynamics of the wing. In addition, the bones of the skeleton

Figure 3.6. A reconstruction of *Hatzegopteryx thambema*, after *Quetzalcoatlus northropii*, a closely related azdarchid pterosaur. Scale = 4m

are lightweight (as in birds), hollow, and thin-walled, yet strong. The small Transylvanian pteranodontid assuredly would have been capable of powered flapping flight that was both agile and maneuverable, but for *Hatzegopteryx*, the situation would have been different. For one thing, the huge head raises the question of how the animal was able to fly at all with such a heavy object stuck on the front of its body. The secret appears to be that the skull's internal bone was constructed of a dense network of very thin trabeculae enclosing small spaces.[12] This polystyrenelike structure would have been exceptionally well designed to combine strength with lightness, of obvious necessity for a flying creature. However, unlike the powered maneuverability of the small Hațeg pterosaur, we expect that *Hatzegopteryx* would have elegantly soared across the sky, silhouetted against the clouds.

Instead of making more specific claims about the form and lifestyle of the Hațeg pterosaurs, we can only form opinions by looking at their closest relatives. When ambling on the ground, pterosaurs appear to have had an awkward, quadrupedal gait,[13] and this condition probably holds for the two Hațeg pterosaurs. In keeping with their presumed warm bloodedness, the body (but not the wings) of these fliers is thought to have been covered with a furlike pelage. The neck may have been long and the tail very short, as in other pterodactyloid pterosaurs. The Hațeg pterosaurs may have had a crest on the front of the upper jaw and perhaps on the undersurface of the lower jaw. Such a crest may have stabilized the head during flight—sort of like sticking the rudder in front instead of behind, as in most present-day airplanes—or it may have had some sort of significance in courtship and territorial displays, much like the cranial ornamentation in hadrosaurid dinosaurs we previously discussed.

What these Hațeg pterosaurs ate is anything but obvious. Based on evidence from other pteranodontid and azhdarchid pterosaurs,[14] the snout was probably long, narrow, and pointed, and the jaws were almost certainly edentulous. The giant pterosaurs may have ventured far out to sea, skimming over the surface of the water to feed on fish. Alternatively, they may have been attracted to a more terrestrial food source, feeding on carrion, such as the carcasses of dinosaurs. In the case of *Hatzegopteryx*— by far the largest predator known from the Hațeg assemblage—it might be at the top of the food chain, the top predator among the area's fauna.

THE HAȚEG MESOEUCROCODILIANS

Turning from the air back to land and water, we now consider the other major group of archosaurs represented in the Hațeg fauna: Mesoeucrocodylia. The remains of at least five crocodilian-related forms have been recovered from Transylvania. Two are ziphodonts (*Doratodon* and an as-yet unnamed form with molariform teeth), whereas another is a short-snouted taxon closely related to *Acynodon* (figure 3.7).[15] The fourth, a newly recognized species named *Theriosuchus sympiestodon*, is another small, short-snouted terrestrial form that appears to be a relict taxon, given its older congeneric relatives from elsewhere in world (figure 3.8).[16] The fifth, *Allodaposuchus precedens* (figure 3.9), is known from a portion of the skull roof, teeth, vertebrae, and limb elements, which were originally collected in 1914 at Vălioara, from the Densuș-Ciula Formation, by Ottokar Kadić, the chief geologist at the Magyar Állami Földtani Intézet, as well as a recently discovered, nearly complete skull collected by members of the joint expedition from Universitatea Babeș-Bolyai Cluj Napoca and the Institut Royal des Sciences Naturelle de Belgique.

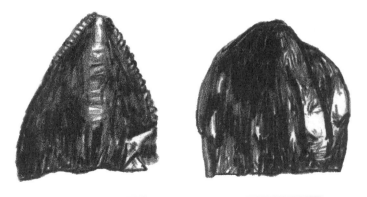

Figure 3.7. The teeth of *Doratodon* (*right*) and a short-snouted taxon closely related to *Acynodon* (*left*). Scale = 1 cm. (After J. Martin et al. 2006)

Figure 3.8. The maxilla of *Theriosuchus sympiestodon* in lateral view. Scale = 1 cm. (After J. Martin et al. 2010)

Nopcsa first described Kadić's material in 1915.[17] Originally not impressed by the details of his new material, Nopcsa considered this Hațeg crocodilian an example of *Crocodilus affulvensis*, otherwise know from the French Late Cretaceous and named by Philippe Matheron.[18] However, by 1928, he had opportunity to examine Matheron's original material and

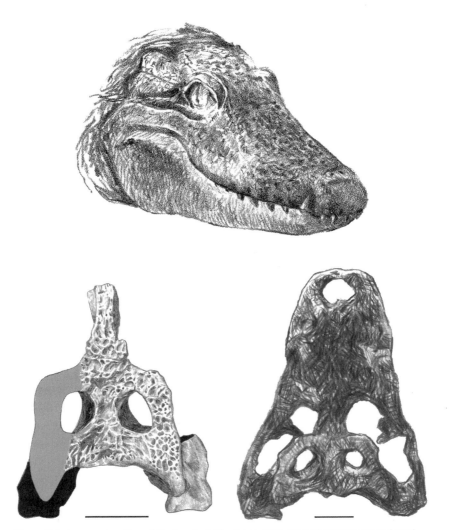

Figure 3.9. A reconstruction of the head of *Allodaposuchus precedens* (*above*), a dorsal view of the skull roof of *A. precedens* (*bottom right*), and the newly discovered skull of *A. precedens* in dorsal view (*bottom left*). Scale = 5 cm. (Newly discovered skull after Delfino et al. 2008)

decided that the Hațeg form was not the same kind of crocodilian, so he provided it with a new name, *Allodaposuchus precedens*.[19]

Based on Nopcsa's original material and a newly discovered skull, we know that *Allodaposuchus* was a medium-sized crocodilian for its time, probably 2 to 2.5 m long.[20] Like most present-day crocodilians, its skull is moderately long, deep, profusely ornamented with pits, and thick-boned. The largest teeth in the lower jaws insert into notches on the sides of the snout, but the number of teeth is fewer than those seen in other crocodilians the size of *Allodaposuchus*. The eye sockets are large and indicate that the eyes and their protective lids stood up from the top of the head. What is known of the rest of the skeleton is limited largely to limb elements that, according to Mason Meers at the University of Tampa, suggest that *Allodaposuchus* was probably a lightly built, mostly terrestrial crocodilian.[21]

In 2001, we joined Angela Buscalioni and Francisco Ortega from the Universidad Autónoma de Madrid to explore the evolutionary relationships of *Allodaposuchus*.[22] According to our cladistic analyses (box 3.2), *Allodaposuchus* appears to occupy a position very close to the base of the group comprising all living crocodilians. A similar position—basal in Eusuchia—was also assigned in a 2008 analysis Delfino et al. made after their discovery, based on a new and nearly complete skull of *Allodaposuchus*, so we can be doubly sure that this basal position is likely to be correct.[23]

Much less is known (from isolated teeth only) about *Acynodon*, another of the Hațeg crocodilians, than about *Allodaposuchus*.[24] These teeth had short crowns that were globular in occlusal view. A recent study of *Acynodon* by Delfino, Martin, and Buffetaut, based on much better material (cranial and postcranial remains from the Santonian of northeastern Italy), indicates that it was a short-faced alligatoroidean.[25] Its jaws were lined with molariform teeth and, unlike many other crocodilians, there was no caniniform tooth. This dentition, as well as aspects of the skull, suggest that *Acynodon* fed on a mixture of invertebrates (mollusks and crustaceans) and possibly plants.

Finally, *Doratodon* from Transylvania is also known only from isolated teeth.[26] Similar to those from elsewhere in Europe, the Transylvanian teeth are compressed transversely, with large serrations on the

BOX 3.2 **Evolutionary Relationships in Mesoeucrocodylia**

Mesoeucrocodylia contains quite a number of fossil forms at its base, as well as Crocodylia, the great clade of all living crocodilians: Gavialoidea, Alligatoroidea, and Crocodyloidea, plus all their extinct relatives. Close to crown-group Crocodylia—indeed, often regarded as its sister taxon—is none other than our own *Allodaposuchus precedens*. *Theriosuchus* and then *Doratodon* have more distant relationships with crown-group Crocodylia and *Allodaposuchus*.

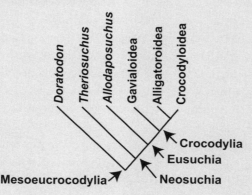

Note: See Brochu 1997, 1999; Buscalioni et al. 2001; Delfino et al. 2005; Pol et al. 2009.

forward and rearward edges. A newly described, nearly complete mandible from Spain indicates that *Doratodon* was relatively small and had a short, narrow rostrum equipped with a reduced, heterodont dentition. On the basis of these and other features, Julio Company and his colleagues considered *Doratodon* to be a mesoeucrocodilian, positioning it as the sibling taxon to Sebecosuchia within Ziphosuchia.[27]

These discoveries suggest that there is greater ecological diversity in Transylvanian taxa on the line toward living crocodilians than had been previously recognized.[28] We suspect that *Allodaposuchus*, the best known among them, was probably an opportunistic, highly efficient predator, preferring to hunt and feed mainly at night, as is true in present-day crocodilian forms.[29] These prehistoric creatures may have tracked their next meal, overpowering and dismembering the body before swallowing chunks of the carcass in single gulps. The ability of crocodilians to eat almost any animal they can overpower or drag underwater and drown has contributed to their success as predators.

The crocodilian habit of basking during daylight hours has led to the popular misconception that these animals are idle predators with limited social acumen. However, present-day crocodilians are known to engage in complex social behavior, including gregariousness, visual and vocal display, territoriality, and courtship behavior. Mothers build and guard nests, and they aid newborns in their struggle to leave their eggs. In addition, parental care extends well into the lives of these hatchlings. Whether a similar behavioral repertoire was present in any of the Transylvanian crocodilians is, unfortunately, still unknown.

As we have noted, the role of the top dinosaurian predator from the Transylvanian assemblages may have been occupied either by small, agile theropods or perhaps by the enormous pterosaur *Hatzegopteryx*. Could it be that *Allodaposuchus* also shared in this predatory biological role? Nopcsa, for one, regarded this ecological status as a distinct possibility. He even suggested that the fragmented, disarticulated dinosaurian remains in places like Sânpetru represented the ancient feeding grounds of *Allodaposuchus*.[30] While we're not so sure about this, it is true that *Allodaposuchus* and its crocodilian cohort would have been formidable predators on whatever they could get a hold of, not only the young and debilitated, but also any of the Transylvanian creatures caught unawares.

THE HAȚEG SQUAMATES

Although archosaurs dominate the Transylvanian assemblages, they do not represent the complete roster of taxa known from this region during the Late Cretaceous. The squamate (lizards and snakes) portion of the Hațeg fauna is known from screening sediment for microfaunal remains. Through these efforts, we know that scincomorphs predominate (box 3.3). Known principally from jaws, teeth, and isolated parts of the skull, Hațeg forms—such as *Becklesius nopcsai*, *B. hoffstetteri*, and *Bicuspidon hatzegiensis*—were probably swift, long-tailed forms resembling their present-day relatives (figure 3.10 on p. 74).[31]

In contrast, anguimorphs represent the slow-moving, secretive predators among present-day lizards. Known only from an upper jaw of an

BOX 3.3 **Evolutionary Relationships in Squamata**

Living snakes, lizards, and amphisbaenians, plus all the descendants of their most recent common ancestor, form the clade known as Squamata (meaning "the scaled ones"). Systematically, squamates include iguanians (agamids, chameleons, and iguanids) and scleroglossans. The latter include anguimorphs (monitors, Gila monsters, and alligator lizards, among others), amphisbaenians (worm lizards), geckos, scincomorphs (skinks, whiptail lizards, and others), and serpents (snakes). The phylogenetic relationships of these taxa are provided in the figure.

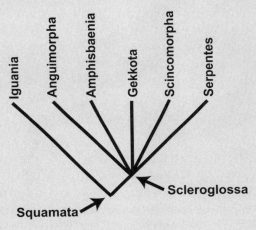

Note: See Vidal and Hedges 2005; Conrad 2008.

indeterminate anguimorph, this clade is very rare in the Hațeg fauna thus far.[32]

Finally, a single vertebra represents all that we know about snakes in the Late Cretaceous of Transylvania. This snake, an indeterminate madtsoiid, apparently killed its prey by constriction, much like present-day boids, to which, however, they are not particularly closely related.[33]

THE HAȚEG TURTLES

Turtles are ubiquitous—scutes, limb elements, and rare skull materials have been collected nearly everywhere, ranging from localities in the Hațeg Basin to Vurpăr, Jibou, Bărăbanț, Lancrăm, Sebeș, and Oarda de

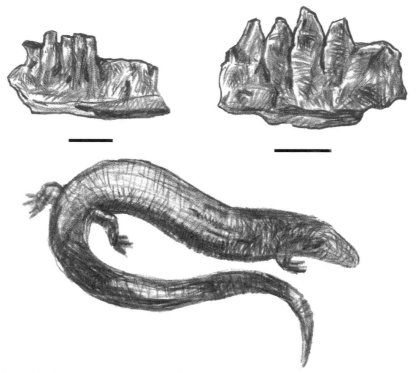

Figure 3.10. The scincomorphs *Becklesius nopcsai (top left)*, *Bicuspidon hatzegiensis (top right)*, and a modern scincomorph, *Chalicides chalicides* (the three-toed skink) *(bottom)*. Scale = 1 mm *(top left)*, 500 μm *(top right)*

Jos in the Transylvanian Depression.[34] One of these has been known for some time. Named *Kallokibotion bajazidi* by Nopcsa in 1923,[35] it may have been quite abundant: fragments of its shell occur everywhere in rocks from the region, although other parts of the skeleton are relatively rare. *Kallokibotion* was a moderately large, broad-headed turtle, with an oblong shell measuring much as 50 cm in length and 40 cm in width (figure 3.11), about the size of a present-day snapping turtle. Like present-day turtles, it was toothless, and the margins of its upper and lower jaws were covered by a horny beak. A slow ambler on land, *Kallokibotion* most likely fed on low-lying foliage and fleshy fruits. Living turtles lay clutches of leathery eggs, which are then abandoned; there is no parental care. We think this may also have been the case with *Kallokibotion*.

Nopcsa regarded *Kallokibotion* as closely related to several primitive

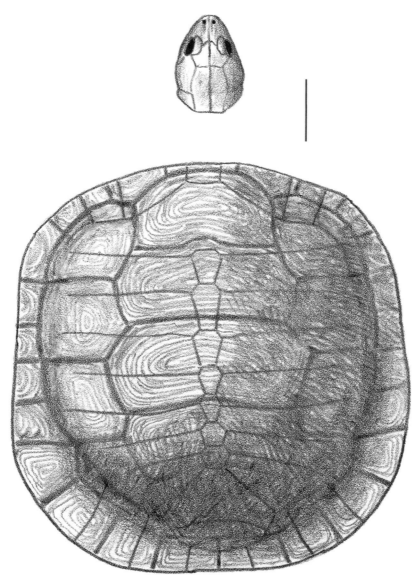

Figure 3.11. A restoration of the skull and carapace of *Kallokibotion bajazidi*. Scale = 5 cm

BOX 3.4 Evolutionary Relationships in Testudinates

Turtles—all living forms and a myriad of fossil taxa—are scientifically known as Testudinata. Characterized by the presence of a shell (composed of a dorsal carapace and a ventral plastron) that completely encloses both of the limb girdles, the absence of teeth lining the jaws, and other features, turtles are known from as far back as the Late Triassic and Early Jurassic, not only from their most primitive taxon, *Proganochelys*, but also from the earliest members of many of today's major groups.

Testudines basally consist of *Proganochelys* and Rhaptochelydia, the latter divided into Austrochelyidae and Casichelydia. Casichelyds make up all modern turtles and their extinct relatives, including Pleurodira (side-necked turtles) and Cryptodira (hidden-necked turtles). The monophyly of these two groups is well supported by both morphological and molecular evidence. Cryptodires make up the greatest diversity among living and extinct turtles. Basally, this group includes the Early Jurassic *Kayentachelys* from Arizona, an unnamed clade consisting of Meiolaniidae and others, and Selmacryptodira. *Kallokibotion* is thought to be the sister taxon of Diacryptodira, consisting of the remainder of all cryptodirans.

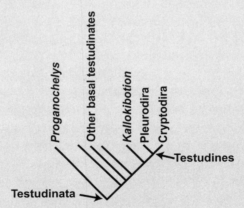

Note: See Hirayama et al. 2000; Joyce 2007.

cryptodiran turtles, those forms that pull their heads into their shells by folding their necks vertically (the other major group of turtles are pleurodires—side-necked turtles—that bend their necks sideways). Recent studies now position it as the sister taxon to Testudines, the crown group consisting of the most recent common ancestor of all cryptodires and pleurodires and all the descendants of this ancestor (box 3.4).[36]

Another turtle—a yet-to-be-determined dortokid—has recently been announced by Vlad Codrea and his coworkers.[37] Previously known from the Cretaceous of southern France and northern Spain, and from the early Tertiary of Romania,[38] they have now been collected from Oarda de Jos and Vurpăr. Dortokids are thought to be primitive pleurodires.

THE HAȚEG MAMMALS

By this time in Earth's prehistory, there was one great mammalian dynasty—Multituberculata—and, emerging from its shadow, two fledgling groups, marsupials and our own group, placentals (box 3.5). The most diverse and abundant of Mesozoic mammals, multituberculates are first known with certainty from Upper Jurassic deposits,[39] but especially so in the Cretaceous fossil record (they survived the Cretaceous–Tertiary mass extinction, flourishing for another 35 million years before dying out in the early Oligocene 30 million years ago). These mammals ranged in size from that of a small mouse (25 g) to that of a woodchuck (4 kg).

Only multituberculates have been found in the Hațeg strata thus far, but it is not known whether the absence of marsupials and placentals is biologically real or whether they will be discovered eventually in Transylvania. The first multituberculate to be described, in 1985, was based on an isolated upper incisor from the Pui locality, and two lower molars from the same site were described in 1986.[40] Known as *Barbatodon transsylvanicus*,[41] the lower molar from this creature was the best material available from the Hațeg Basin until 1996, when the first multituberculate skull from Sânpetru—named *Kogaionon ungureanui*—was described by two paleontologists from Institutul speleologic "Emil Racoviță" București, Costin Rădulescu and Petre-Mihai Samson (figure 3.12).[42] A possible new species of *Kogaionon* was reported by Vlad Codrea and his coworkers in 2002.[43] Finally, a multituberculate called *Hainina*, otherwise known from Paleocene rocks in western Europe, was also discovered in 2002 from the Late Cretaceous of the Hațeg Basin.[44]

BOX 3.5 **Evolutionary Relationships in Mammalia**

Mammalia, the large group of vertebrates in which we claim membership, presently dominates the world's biota of large animals. The northern high latitudes have their musk ox, elk, and polar bears; the tropics, their tapir, giraffe, elephants, rhinos, hippos, tigers, and lions. Placentals all, they are united by having a special construction of the embryonic membranes between mother and embryo called the placenta. By comparison, marsupials have only an incomplete placenta, but most females carry their young in an abdominal pouch called a marsupium. Today, marsupials are best known from Australasia, but they are also found in South America, and—thanks to the pesky Virginia opossum (*Didelphis virginiana*)—in North America as well. Together, placentals and marsupials are grouped as Theria. The remaining living mammal group, the egg-laying Monotremata, includes platypuses and echidnas from Australia and New Guinea.

Because all these mammalian groups are still living, we consider them crown-group Mammalia, but they alone aren't all of the major groups represented in this clade. Another group, Multituberculata, is more closely related to Theria than Monotremata is. First appearing in the Late Jurassic and going extinct in the Oligocene, these small herbivores are well represented all over the Northern Hemisphere.

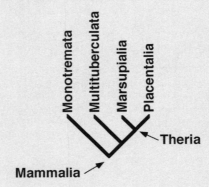

Note: See Kielan-Jaworowska and Hurum 2001; Kielan-Jaworowska et al. 2004; Lillegraven et al. 1979; McKenna and Bell 2000.

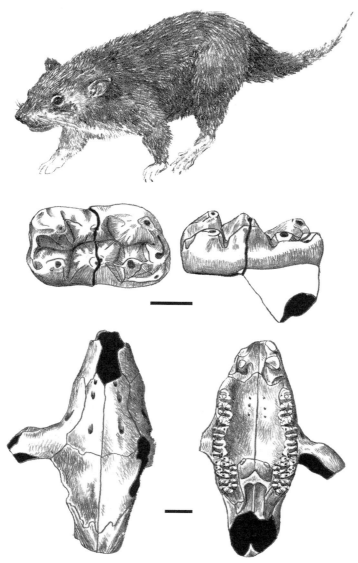

Figure 3.12. A reconstruction of *Kogaionon ungureanui* (*top*), and the lower left first molar of the Hațeg multituberculate *Barbatodon transsylvanicus*, in occlusal (*middle left*) and lingual (*middle right*) views. Scale = 1 mm. Dorsal (*bottom left*) and ventral (*bottom right*) views of the skull of the Hațeg multituberculate *Kogaionon ungureanui*. Scale = 5 mm. (Top and middle after Rădulescu and Samson 1986; bottom after Rădulescu and Samson 1996)

BOX 3.6 **Evolutionary Relationships in Amphibia**

Modern amphibians (Lissamphibia) include frogs and toads (Anura), salamanders and newts (Caudata, or Urodela), and caecilians (Gymnophiona, or Apoda). In addition, a number of extinct taxa are known within Lissamphibia. As concerns us here, Albanerpetontidae is known from the Jurassic to the Miocene. Members of this wholly extinct clade most closely resemble modern salamanders.

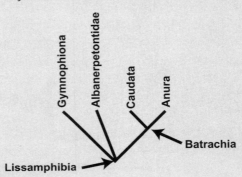

Note: For Gymnophiona, see Frost et al. 2006. For Albanerpetontidae, see Gardner 2000, 2002; Gardner et al. 2003; McGowan 2002; Venczel and Gardner 2005.

Multituberculates have a long, broad skull that is superficially rodent-like in appearance, with long lower incisors, followed by a gap (called a diastema) between them and the cheek teeth. The lower premolars are specialized into shearing blades, whereas the molars are designed for crushing or grinding food, suggesting that multituberculates, including those from Hațeg, had a varied diet that included seeds as well as other larger, hard food items.[45]

Based on features of the most complete skeletons, paleontologists believe that at least some multituberculates were arboreal.[46] With a prehensile tail, a hind limb held in a crouched position, a reversible ankle similar to that seen in squirrels, and a wide range of forelimb motion, these forms could have descended tree trunks headfirst to forage on the ground. However, other multituberculates may have been burrowers, and a third interpretation is that some forms were leapers.[47] Whatever the case (and these interpretations are not mutually exclusive), it would be premature

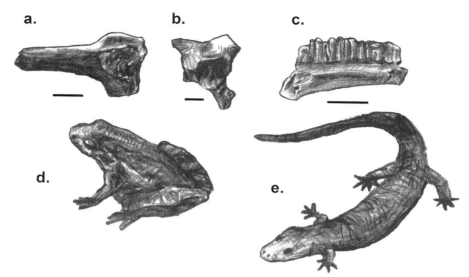

Figure 3.13. The dentaries (with teeth) of (a) *Hatzegobatrachus grigorescui* (an anuran), (b) *Paralatonia transylvanica* (an anuran), and (c) *Albanerpeton* sp. Scale = 0.5 mm. A living anuran, (d) *Discoglossus galganoi*, and (e) a reconstruction of the extinct *Albanerpeton* sp.

to characterize the Transylvanian forms as leapers, burrowers, or tree climbers without any postcranial skeletal material. Whatever their mode of locomotion, *Barbatodon transsylvanicus* and *Kogaionon ungureanui* would probably have been found scurrying by moonlight in the mosaic of shadowed alcoves, branches, or undergrowth, as did many nocturnal Mesozoic mammals, since the best escape from predatory dinosaurs was the cover of darkness.

THE HAȚEG AMPHIBIANS

Active screening for microvertebrate remains in Transylvania in the past two decades has produced a number of new frogs and an albanerpetontid (box 3.6).[48]

Hatzegobatrachus grigorescui and *Paralatonia transylvanicus* are small frogs (with an estimated 30–50 mm snout–vent length) known from jaw elements and various postcranial elements. Other indeterminate frogs have been reported from the Hațeg Basin: one that appears to be closely related to *Eodiscoglossus*, and another having an affinity with *Paradiscoglossus* (figure 3.13).[49]

BOX 3.7 **Evolutionary Relationships in Actinopterygii**

Actinopterygii, also known as ray-finned fish, possess fins supported by lepidotrichia (bony or horny spines). Currently consisting of nearly 30,000 species, actinopterygians are found in both freshwater and marine environments, from the highest mountain streams to the deepest sea. Three groups of actinopterygians (acipenseriforms, lepisteostids, and teleosts) are recognized from the Late Cretaceous of Transylvania.

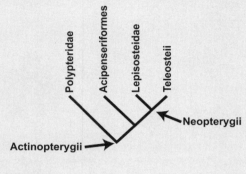

Note: See Stiassny et al. 1996.

Albanerpeton, an enigmatic, extinct, salamanderlike taxon, is the last of the amphibians recovered from the Late Cretaceous of western Romania.[50] Otherwise known from the Middle Jurassic to the Miocene of Europe, North America, and central Asia, albanerpetontids were approximately 10 cm long, predominantly terrestrial (although some may have been fossorial), and predatory.[51] Albanerpetontids are placed within Lissamphibia (the crown group that includes frogs, salamanders, and caecilians).[52]

THE HAȚEG ACTINOPTERYGIAN FISH

The Hațeg fauna, despite coming from sediments laid down in streams and rivers (chapter 4), is counterintuitively very poor in fish of any kind (box 3.7). Thus far, the only known fossil remains from members of this fully aquatic community are acipenseriforms (chondrosteans), lepisteostids (holosteans), and teleosts. Living acipenseriforms consist of sturgeons and paddlefish, both with cartilaginous skeletons, elongate bodies,

Figure 3.14. Living representatives of the Transylvanian actinopterygians: gar (*Lepisosteus oculatus*; *top*); blue neon tetra (*Paracheirodon simulans*; *middle*), a characiform teleost; and sturgeon (*Acipenser transmontanus*; *bottom*)

and triangular snouts. Acipenseriforms are bottom feeders. Rhomboidal, thick scales and a conical, slightly recurved tooth attest to the Haţeg presence of the gar *Lepisosteus*.[53] Present-day gars are long-snouted carnivores, with a long, slender body and a heterocercal caudal fin. Finally, characiform teleosts have been reported from the Haţeg Basin.[54] Today we know some of these characiforms as the popular aquarium pets called tetras (figure 3.14).

CHAPTER 4

Living on the Edge

What was the Transylvanian region like some 70 million years ago, during the Maastrichtian age of the Late Cretaceous? To Franz Baron Nopcsa, the prehistoric landscape had a distinctly different appearance than it does in modern times. Here is how he might have imagined it, with a bit of a current paleogeographic spin.

Seen from high in the air, the Transylvanian region is an island—a speck among other specks—that, together with its neighbors, forms a chain that extends across the northern flank of a broad seaway. This great body of water, called the Neotethyan Ocean, covers what is now most of Europe and Russia to the north and northern Africa to the south. Some islands are small oceanic volcanoes no more than 100 square km in area, but a few are the size of present-day Madagascar. One of the latter, which would form what is now part of southern France and the majority of Spain, is the westernmost island of the chain, some 2,500 km away from Hațeg Island. The other end of the chain, in what is now the Caucasus, lies 2,000 km away from us, and we are some 3,000 km from the nearest coast of northern Africa (Plate VI).

Thus our island of interest—Hațeg Island—is halfway along this chain, some 20°–30° North Paleolatitude, just below the Late Cretaceous position of the Tropic of Cancer. As we descend through the tropical clouds toward our island, the air becomes more humid and we see several high-flying birds and pterosaurs. The green flanks of volcanoes stand out against the deep blue of the surrounding Neotethyan Ocean as we begin to perceive the horseshoe shape of the island and its extent, some 75,000 square km in area (about the size of present-day Sri Lanka).[1]

As we get ever closer, the neighboring islands disappear from view. An occasional ash cloud drifts up from the cauldron of one of the volcanoes on Haţeg Island, but otherwise the air is clear—clear and thick with humidity and the white noise of insect cacophony. The climate on Haţeg Island is largely controlled by the surrounding warm currents of the Neotethyan Ocean and the prevailing westerly winds. Parts of the island receive abundant but seasonal rains, such that the island undergoes alternating wet and dry periods.

The island's details emerge and we see not a single geometric unit, but tiers of diverse habitats and their denizens. Beaches ring the periphery; the island is rugged in places, and lava hints at a violent volcanic past. Most of the lower elevations of the island are covered with diverse rain forests. Farther inland, a series of low hills and terraces leads away from the shore. This gently sloping alluvial plain is well drained by numerous, rapidly flowing, shallow rivers. Their channels are interspersed with sand bars and occasional, short-lived, shallow lakes. Abundant fetid swamps occupy the lower-lying regions. Floodplains, on which soils have formed and vegetation flourishes, are broadly distributed among water and wetlands. There is luxuriant plant growth everywhere: dense forests towering above a lush tropical understory, and river banks covered with a riot of colors from flowering angiosperms.

The uplands of the island are often cloudy and humid, although where there is fresh lava, the gray rock is barren and sunbaked. Yet even these highlands have stands of woodland, though not nearly as dense or as common as in the lowlands. Instead, shrubs, weedy plants, and other ground cover dominate the landscape.

NOPCSA, DWARFS, AND ISLANDS

Lecturing at the 27 November 1912 meeting of the Zoological and Botanical Society in Vienna, Nopcsa summarized what was then known about the dinosaurs from the Upper Cretaceous of Transylvania.[2] He then described their evolutionary relationships within Dinosauria, the conditions under which their remains were found, and their paleoecology, which he regarded as swamp dwelling. In doing so, he emphasized that, compared with the dinosaurs of the same age elsewhere in the world, all of the dinosaurs of Transylvania were strikingly smaller:

> During the time of the Upper Cretaceous [Haţeg] formed an island and *the smallness of the Transylvanian Cretaceous dinosaurs can be examined as a consequence of insular life* [italics in original], which in other groups, as for example in the fossil elephants of the Mediterranean islands, likewise led to a reduction in body size in comparison with the continental forms.[3]

Audience members at this lecture were impressed with Nopcsa's inference that the members of the Haţeg fauna were dwarfs, but they were somewhat less than sanguine about whether the dinosaurs were dwarfs because of their island habitation. Pointing out that this was the first time that the Transylvanian dinosaurs had been clearly identified as a dwarfed fauna, Othenio Abel (figure 4.1), the session's convener, objected to Nopcsa's insular hypothesis, predicating his critique on three arguments.[4] First, Abel asserted that the basis for dwarfing could not be directly attributed to island life per se, as was found in the dwarf elephants of Malta and the dwarf hippos of the Mediterranean islands. Instead, he argued that their diminution in size was due to the inevitable inbreeding that comes with insular isolation, as well as through other geographical interventions, such as isolation caused by mountain uplift and river dynamics. As a second objection, Abel noted that insular habitation does not always lead to dwarfing, pointing to examples in which certain reptiles and birds have tended toward gigantism through isolation on islands (e.g., crocodiles in Madagascar, Komodo dragons in Indonesia, turtles from the Galápagos Islands, moas in New Zealand). Third, Abel worried about Nopcsa's use of the word "island" in reconstructing the Haţeg paleoenvironment, questioning whether the region

Figure 4.1. Othenio Abel (1875–1946)

that supported the fauna was actually small enough to be considered an island, or instead was sufficiently large to be considered a mainland.

Abel's comments stayed with Nopcsa as he assembled his case for dwarfed Transylvanian dinosaurs. In 1915, he provided further documentation of body-size differences between the animals he had recovered from the Hațeg Basin and those elsewhere, in order to answer Abel's criticisms.[5] While suggesting that dwarfing either was due to the Transylvanian dinosaurs retaining a primitive stage of evolution or was a symptom of evolutionary degeneration, Nopcsa still appears to have been uncertain about the importance of either. As a good neo-Lamarckian, he explicitly advocated degeneration produced by insular isolation, opining that growth in the bones from Transylvania may have been pathologically limited through illnesses (caused perhaps by malnutrition or starvation).[6] He called these changes in body size—from dwarfing in the Transylvanian dinosaurs to gigantism among dinosaurs in general—arrostic (literally, "arresting") alterations.[7] For Nopcsa, arrostic changes were nothing more than the heritable acquisition of endocrine disease: hyperfunction of the pituitary gland in the evolution of gigantism, hypofunction of the thyroid gland in the evolution of dwarfing. We will return to these biological explanations in chapter 5.

Nopcsa used local geology to explore the paleogeography of the region that would have spurred the evolution of increasingly smaller descendants from larger ancestors. To do so, he took to the rocks, returning to the research he conducted previously while a graduate student at the University of Vienna under the supervision of Eduard Suess.[8] Nopcsa's dissertation research had taken him back home to Transylvania, in the heart of the Southern Carpathians, to conduct a comprehensive and detailed study of the geologic history of the hills, mountains, and rolling plains that he knew so well as a child living in Săcel. In the valleys of the Sibișel, Cerna, and Jiu rivers, the slopes of the Retezat, Sebeș, and Poiana-Ruscă mountain ranges, and the villages of Zeicani, Fărcădin, and Pui, he occupied himself with the real grunt work of a field geologist —walking, sample collecting, note taking, and mapping—as well as the marveling, wondering, and questioning that comes from these physical labors. Nopcsa not only looked at the rocks from oldest to youngest— from the granites, schists, and gneisses dating back to the Precambrian through the overlying sediments dating from the Early Jurassic to the last Ice Age—he also interpreted the ancient environment of depositions and their significance with respect to Earth's major upheavals.

From Nopcsa's detailed dissertation research, published in 1905 as a nearly 200-page monograph, we learn that the Hațeg region 150 million years ago was generally a warm, marine-dominated environment.[9] The widespread limestones and chalk marls that dominate the Lower Jurassic and Lower Cretaceous were precipitated in the warm Neotethyan Ocean. Nopcsa thought that the central part of the Retezat Mountains, whose uplift he regarded as beginning no later than the Early Jurassic, may already have risen above the sea. Had the Hațeg region been an island in the Late Jurassic and Early Cretaceous? Nopcsa, thus far, remained uncertain. Sedimentary evidence of beaches that would have ringed the island—key to his insular hypothesis—is absent on the flanks of these mountains (though Nopcsa attributed this lack of evidence to erosion occurring since the Jurassic). However, following its emergence during the mid-Cretaceous, the Hațeg region was again covered by a warm sea, its waters abounding with predatory ammonites and the strange, reef-building bivalves called rudists, as well as oysters, clams, snails, and single-celled foraminifera. Some evidence of this remains in the region. Sea level fluctuated greatly through this interval, peaking some 80 million years ago, during the Campanian.

Terrestrial conditions at the very end of the Cretaceous are well preserved in the Transylvanian region. Outcrops of these Maastrichtian rocks are known in the Hațeg Basin itself, as well as in the southern parts of the Poiana-Ruscă Mountains and in the Apuseni Mountains (Plates VII and VIII). Nopcsa interpreted these rocks (a thick sequence of conglomerates, sandstones, and mudstones) as a system of nearshore, fluvial, and lacustrine environments, finding support for his interpretation of terrestrial conditions in the unionid clams, turtles, crocodilians, and dinosaurs often found in abundance at various Transylvanian locations.[10] This emergent habitat endured until the early Eocene, approximately 55 million years ago. Sea level rose again in the middle Eocene, as the shales interbedded with coarse limestones overlying the terrestrial deposits testify, only to recede again in the Oligocene.

Nopcsa's last word on the paleogeography of the Hațeg Basin came in 1923.[11] Using the paleogeographic maps available to him at that time, he described the area as a "mere archipelago" of five major islands, distinctly separated from each other and the mainland since the beginning of the Late Cretaceous. Extensive overthrusting, which began some 80 million years ago, produced the present-day Carpathian ranges and the main outlines of current European topography. As Earth's crust buckled, the seas withdrew, uncovering what were to become the Southern Carpathian Mountains. Here, the Transylvanian dinosaurs and other members of the fauna were entombed in the region's riverbeds, freshwater lakes, and formidable bogs.

PLATE TECTONICS AND NEOTETHYAN PALEOGEOGRAPHY

Nopcsa's schooling at the hands of Eduard Suess introduced him to the possibility that the global positions of the continents may not have been fixed over geologic time. In his nineteenth-century attempts to explain the formation of folded mountains ranges like the Alps, Suess suggested that such major geologic upheavals were due to the gradual contraction of the planet.[12] With this shrinking, due to Earth's cooling, its outer crust was forced to wrinkle, fold, and subside. In so doing, large regions of the crust created depressions into which the seas drained, thereby exposing regions of dry land. As Suess's student, Nopcsa came to see Earth as a changeable, mobile matrix of continents and oceanic basins formed through shrinking, rather than as a permanently fixed geography.

Nevertheless, Suess's contractionist theory didn't survive long into the twentieth century before it collapsed—there were just too many improbable estimates of cooling and contraction rates, a recognition that continental and oceanic crusts were different in their composition and density, and the discovery that radioactivity in the crust created a stable heat balance across geologic time. Yet Suess had opened the door widely for the possibility of large-scale crustal mobility.

Although he was not the first to propose major lateral displacements of the continents, one person has been rightly credited as the principal developer of the ideas of continental drift, Alfred L. Wegner.[13] In 1912, Wegener proposed that, in the distant past, all the continents had been united, had later broken apart, and thereafter drifted through the ocean floor to their current locations. Much of his evidence came from the jigsaw fit between the margins of the continents, biogeography, paleoclimatology, and evidence that the present-day continents are actually, though very slowly, moving. Nopcsa himself contributed to Wegener's theory by incorporating aspects not only of paleontology and biogeography, but also of geochemistry and theoretical geophysics.[14] For their time, Nopcsa's arguments for continental mobility—involving the relationship of different kinds of volcanism that came from the movement of continental plates over the top of zones of subducting oceanic crust—were quite innovative, agreeing with the present-day view of magma origin.[15] Nevertheless, many felt that the evidence in support of Wegener's multifaceted theory, called continental drift, was ambiguous, and it failed to attract a powerful following—that is, until the late 1950s and the 1960s.

In the 1950s, geophysicists initiated studies of Earth's past magnetic field, as recorded in geologic deposits, and this work later established considerable variation in the position of Magnetic North through time.[16] In particular, the pole position on one continental landmass at any given time in the past was not the same as the pole positions on the others; instead, there were many pole positions. One explanation for this apparent polar wandering was continental drift. By moving the continents across the face of the globe, the pole positions could all become superimposed. Drift received a spark of renewed interest.

In addition, thanks in part to research spawned by the events of World War II, the oceanic basins were investigated as never before. The discovery of oceanic ridges, trenches, volcanoes, and mountains came from mapping the ocean floor using radar, which imparted considerably more topography than had ever been expected before. Abundant shallow-earthquake activity and volcanism were associated with the ridges, whereas deep earthquakes were linked with the trenches. Soon it was discovered that the age of the oceanic crust was not uniform. By mapping, sampling, and radiometrically dating the oceanic crust, it was revealed that the youngest oceanic floor was at the midoceanic ridges, and the oldest was in proximity to the continents. New oceanic crust was formed at the ridges and spread away laterally, indicating that the continents must have been closer to each other in the past. In other words, they had to have moved to their present positions.

As these new data flooded into laboratories during the 1960s, Wegener's theory of continental drift was to receive a vigorous dusting off.

Wegener's early evidence and all of the more recent discoveries could be causally related in what is now known as plate tectonics. Developed in further detail by geologists, geochemists, geophysicists, and paleontologists, this unified theory links all of the dynamic processes of the globe— how mountains are formed, why volcanoes are distributed in the Ring of Fire around the Pacific margin, how the sea floor is formed, and why certain zones are earthquake prone—with the relative motions of a mosaic of large, rigid, lithospheric plates that constitute Earth's outermost shell. A variety of motions can occur at the boundaries between these plates. Lithospheric plates moving toward each other produce a zone of convergence, while in moving away from each other they form zones of divergence. Plates can also slide past each other to produce transformational movement, such as that along the San Andreas fault system in California. New plate material is produced at divergent boundaries (e.g., along the crests of midoceanic ridges), while old material is destroyed and recycled by subduction at convergent boundaries, as is the case along the deepsea trenches off the Pacific coast of Japan. Given the available evidence, geologists think that each lithospheric plate floats on Earth's upper mantle, a 660 km thick, plastic layer of dense, semimolten, magnesium- and iron-rich silicates. Plate motion is driven by large-scale convection currents in this upper mantle. It is this motion, imparted to the continents, which produced the global patterns that originally caught Wegener's eye.

With the acceptance of plate tectonics in the late 1960s, scientists clamored to reconstruct Earth's geography in bygone times, and in doing so they produced a flurry of paleogeographic maps. At first, these were done by hand, but with the coincident increase in computer access, computer modeling of plate motion and continental repositioning became the working tool of choice. The global models provided in 1970 by Robert S. Dietz and John C. Holden were based on the best fit of the continental positions of present-day coastlines, the margins of the continental slope, and patterns of fracture zones in the ocean floor.[17] Later computer simulations, such as those by Alan G. Smith and his collaborators from England, followed suit, but they added further constraints (figure 4.2).[18] First, based on their present coastlines and shelf margins, the continents were reassembled into their positions prior to the opening of the

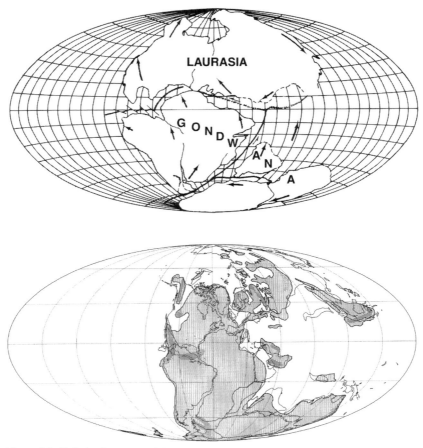

Figure 4.2. Global paleogeographic reconstructions for the Triassic, approximately 220 million years ago. (Dietz and Holden 1970 reconstruction [*above*]; A. Smith et al. 1994 reconstruction [*below*])

Atlantic, Indian, and other oceans, much like the jigsaw-puzzle approach used by Dietz and Holden. Second, a paleogeographic latitude–longitude grid was superimposed on the reassembled continents, and successive stages of breakup—with reference to the positions of the paleomagnetic poles for each interval of time (from those apparent polar-wandering curves)—were determined.[19] It is through such paleogeographic reconstructions that we came to know of the existence of the late Paleozoic supercontinent called Pangea, with its huge eastward-opening Tethyan

Ocean that provided Pangea with a northern Laurasian half and a southern Gondwanan half. This kind of research has also been used to graphically document the opening of the North Atlantic in the Late Triassic–Early Jurassic, and then the South Atlantic in the mid-Cretaceous.

As these computer models of plate tectonics became available, other scientists began investigating more fine-grained changes in global conditions. Most important have been new models of paleoclimatologic change, including variations in oceanic circulation patterns, fluctuations in the surface temperature of the oceans and the atmospheric temperature, changes in patterns of atmospheric pressure and circulation, and alterations in weather patterns.[20] Data taken from depositional environments (based on surface outcrops and boreholes), as well as information on fossil plants thought to indicate particular environments, can also be plotted on paleogeographic maps to characterize important aspects of regional paleoenvironments.

With both these new global and more regional paleogeographic models, we have come to appreciate that the history of the Tethyan region was much more complicated than previously thought. Instead of a single oceanic basin, it was formed by two different, spreading ridge systems, and therefore had two successive basins—the Paleotethys and the Neotethys. By the Late Jurassic, the Paleotethyan Ocean, the great incisure into Pangea that had formed in the Devonian, was on its way out as the Neotethyan Ocean increased at its expense throughout the rest of the Mesozoic. The Paleotethys formed in the Devonian, when a stretch of European terranes (which were to become northern Europe) separated northward from the northern margin of what is now known as Africa. This opening ocean, created by the Paleotethys midoceanic spreading ridge, gave the world's geography a huge incisure that opened to the east and partially separated Pangea into Laurasia and Gondwana for much of the remaining Paleozoic. The Neotethyan Ocean began to close toward the east as the African plate rotated counterclockwise and the northern margin of the Neotethyan Basin began undergoing subduction. This produced a number of island arcs in what is now the region of the Middle East and the Caucasus.[21] In addition, subduction of the Neotethyan oceanic crust under the more southerly Arabian portion of the African continental plate formed an island chain across present-day Iran.[22]

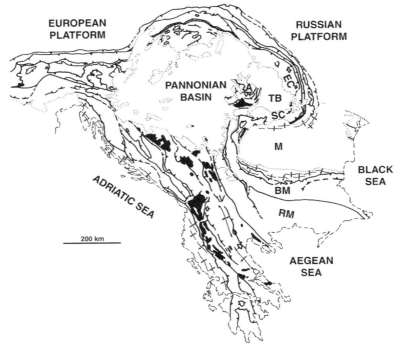

Figure 4.3. A generalized tectonic map of the Eastern European Alpine system. A: Apuseni Mountains; BM: Balkan Mountains; EC: Eastern Carpathian Mountains; M: Moesian Platform; RM: Rhodopian Mountains; SC: Southern Carpathian Mountains; TB: Transylvanian Basin. (After Burchfiel 1980)

In order to understand how these Neotethyan tectonics relate to the Transylvanian region, we now need to explore the history of what today are the Western, Eastern, and Southern Carpathian Mountains, that great mountainous loop crossing Slovakia, Ukraine, and western Romania (figure 4.3). The history of these mountain belts is very messy, to say the least. The region has seen times of incredible deformation, severe faulting, great compressional forces, crustal warping, zones of extension and collision, and the suturing of continental fragments against the irregular European-Russian platform. These events took place throughout the Cretaceous, and they were subsequently highly modified by further regional deformation during the Cenozoic.[23] Research by Mircea Săndulescu (Universitatea din București), by B. Clark Burchfiel (Yale

University), and, most recently, by Ernst Willingshofer (Vrije Universiteit Amsterdam) has deciphered a good deal of the complex paleogeographic history of this region, thus providing us with a view of the events that gave the Transylvanian dinosaurs a place to live.[24]

By the Late Jurassic, numerous fragments of continental crust lay immediately south of the European platform in the northern Tethyan realm (figure 4.4). Two of these fragments—Apulia and Rhodope[25]—are of particular importance to us. Here the margin of the European platform was very irregular, with a substantial promontory—the Moesian platform[26]—extending westward to form an important northern embayment between it and the European plate.[27] This embayment has its origin in the opening of the Alp-Tethys oceanic basin in the Early Jurassic. At that time, Rhodope lay to the west and south of Moesia. The short-lived Severin Ocean separated Rhodope and Moesia, while the Vardar Ocean was to the south and west of Rhodope, separating it from Apulia. Both of these oceans were relatively sizeable deepwater domains, with Vardar larger than Severin. On Apulia, shallow-water carbonates dominated, but there also appear to have been regions of exposed land.[28] Likewise, Rhodope consisted of a mixture of one or more exposed islands, shallow water carbonates, silts, and sands. In contrast, Moesia and adjacent parts of the European platform appear to have been covered by shallow seas.[29]

From the Late Jurassic onward, the kinematics of convergence of these microplates ultimately controlled by movements of the African platform relative to the European platform, has Rhodope beginning its north- and eastward migration toward Moesia, thereby closing the Severin Ocean sometime during the Aptian, about 120 million years ago. By the Early-to-Late Cretaceous transition (approximately 105–90 million years ago), the Severin Basin had closed and the Rhodope microplate began docking with Moesia, a process that continued for another 15 million years, into the Campanian. By that time, having wrapped itself around the western margin of the Moesian promontory, Rhodope was shifting to more continental conditions.

The northward and eastward movement of the Apulian fragment toward Rhodope began in earnest early in the Late Cretaceous and by the Campanian–Maastrichtian interval, some 80–70 million years ago,

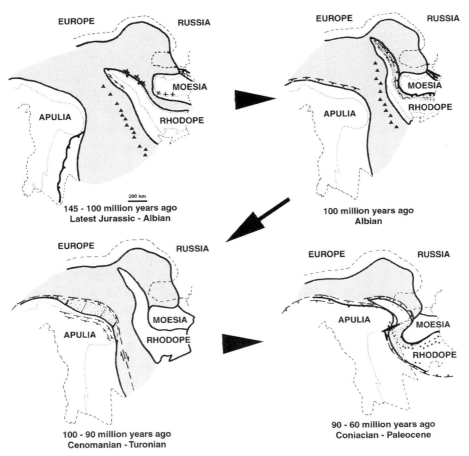

Figure 4.4. Relative positions of the Apulian, Rhodopean, and Moesian microplates from the latest Jurassic through the beginning of the Tertiary. The shaded regions represent the possible extent of the oceanic crust; triangles are suggested locations of volcanism; barbed lines indicate subduction zones; the dashed line at the top of each diagram indicates the present location of the Carpathian–Balkan front; the dashed line around the boundaries of the microplates are present-day boundaries along the Black, Aegean, and Adriatic seas; the dotted lines are the present sizes of microplate boundaries. (After Burchfiel 1980)

Apulia had begun its docking with Rhodope. As this migration of Apulia continued during the Paleocene–Miocene (65–15 million years ago), it induced the clockwise rotation of Rhodope, which had the effect of wrapping Rhodope completely around the Moesian promontory into the European embayment. Various parts of the Apulian plate thrust north-

and eastward to occupy the wedge between Rhodope, Moesia, and the European platform and add to the continental conditions of the Transylvanian region. It is through all of these collisions, which began in the Cretaceous and continued through much of the Tertiary, that the Western, Eastern, and Southern Carpathian mountain belts were formed (figure 4.3). As it was shoved against the margins of Moesia, Rhodopian tectonics formed the Southern Carpathians—which includes the Haţeg Basin—and small portions of the adjacent Eastern Carpathians. The remainder of the Eastern and all of the Western Carpathians are due to the collision and suturing of Apulia with the European platform in the Tertiary. The Apuseni Mountains—the heart of Transylvania, lodged between the Southern and Eastern Carpathians—are also the tectonic result of the forward regions of Apulia pushing north- and eastward toward the European plate.

Not only were continents adrift and new oceanic basins forming in the Neotethyan realm during the Cretaceous, but these great crustal dynamics also had profound effects on changes in the sea level, patterns of oceanic circulation, and the climate, all of which influence our understanding of the conditions faced by the dinosaurs of Transylvania. Sea level in Europe (and for much of the rest of the globe) rose and fell throughout much of the Early and Late Cretaceous.[30] The largest of the Late Cretaceous transgressions began about 100 million years ago, and it is thought to be related to an increase in rates of seafloor spreading and the addition of new oceanic ridge systems.[31] It certainly reduced the entire European coastline dramatically, allowing the Neotethyan Ocean to communicate not only with the North Sea, but also, through the Polish Trough and the Paris Basin, with the North Atlantic (Plate VI). Flooding of what is now the region of Ireland and England, and much of France, Spain, and central and eastern Europe, culminated approximately 80 million years ago (during the Campanian) and produced the greatest number of European islands, both large and small. To the north and west was the largest, Fennosarmatia (present-day Fennoscandia and northern European Russia). In the extreme west and south lay several smaller islands, including the Iberian Meseta (Portugal and western Spain), the Ebro High (northeastern Spain), and the Massif Central (southcentral France). Several other exposed regions are located farther north and closer to Fennosarmatia (including the Cornwall region of southwestern England, the Irish Massif, and Caledonia). Finally, the Rhenish Massif (northeastern France and southern Germany) occupied a more central European realm.

NO PLACE LIKE HOME

Since Nopcsa's time, we have learned a great deal more about the paleogeography of Transylvania during the Late Cretaceous. The Haţeg Basin is an intramontane basin measuring approximately 45 km from east to west, bounded by the Şureanu Mountains to the northeast, the Poiana-Ruscă Mountains to the northwest, and the Retezat Mountains to the south (figure 4.5). Narrow zones of crystalline basement rock and marine sediments separate the Haţeg Basin from the Petroşani and Rusca Montană intramontane basins. The Petroşani Basin to the east contains Oligocene and Lower Miocene freshwater lake and brackish water deposits, while the Rusca Montană Basin to the west contains Upper Cretaceous marine and fluviolacustrine deposits, preserving abundant coal beds and a reasonably well-known terrestrial flora. Finally, the Transylvanian Depression (also known as the Transylvanian Basin[32])—a very large, subcircular basin to the north and northeast of the Haţeg Basin, within the bend zone connecting the Eastern and Southern Carpathian Mountains—hosts more than 10 km of sediment that date from the Late Cretaceous through the Pliocene.

During the latest Cretaceous, the Haţeg Basin was located on the eastern flank of the Rhodope microplate, whereas the localities of the adjacent Transylvanian Depression were scattered around the forward and northern edge of the Apulian microplate. At that time, all were located about 20°–30° North Paleolatitude.[33] Surface currents in the nearby marine realm were probably westward, and surface water temperatures may have averaged no less than 24°C in the winter and 32°C in the summer.[34] This part of the Neotethyan Ocean may also have been susceptible to violent storms.[35] A diverse array of predatory ammonites and a host of single-celled foraminifera lived in these waters, but the marine vertebrate biota of Transylvanian is unknown. In nearby regions, however, teleost fish, sharks, marine crocodilians, and mosasaurs are present.[36]

The surrounding warm waters of the Neotethyan Ocean and the prevailing westerly winds most likely made the terrestrial climate of the

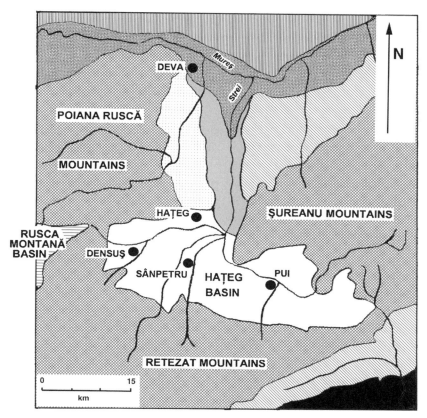

Figure 4.5. Major geographic features around the Haţeg Basin. (After Weishampel et al. 1991)

Transylvanian region warm and humid, at least seasonally. In addition, what today are the Retezat and Poiana-Ruscă mountains were beginning to form, providing the principal topography of the Transylvanian landscape with a source of sediments during the latest Cretaceous.

The Transylvanian dinosaur fauna comes from strata exposed in numerous localities within the Haţeg Basin and the Transylvanian Depression (figure 4.6). Thus far, research on the Haţeg Basin is more extensive and detailed than that on the Transylvanian Depression, so the former will be covered first.

The Haţeg Basin includes two distinctly different sequences of rock, which, as a consequence, represent two formational units: the Sânpetru

Living on the Edge

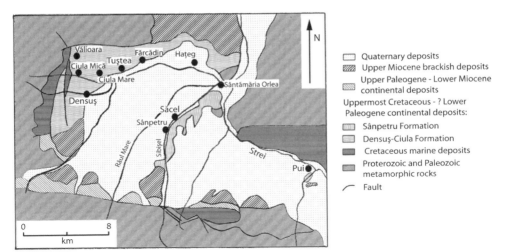

Figure 4.6. A geologic map of the western part of the Haţeg Basin. (After Weishampel et al. 1991)

Formation outcrops in the central part of the southern end of the basin, and the Densuş-Ciula Formation in the west (Plate VIII). The Sânpetru Formation—a rock unit thought to be an outcropping at least 2,500 m thick on both sides of the Sibişel Valley, along the Râul Mare all the way to Sântămărie Orlea, and along the Râul Bărbat at Pui—has yielded the richest lode of fossil vertebrates from the Haţeg Basin.[37] Sedimentological research indicates that the Sânpetru Formation consists almost exclusively of clastic sediments, from coarse conglomerates (including pebbles 15–20 cm in diameter—about the size of large grapefruits) to claystones.[38] The Formation also includes irregular beds of calcium carbonate nodules. Toward the top of the Sânpetru Formation, strata not only include the finer-grained beds (now dark gray or green instead of brown or red[39]), as seen lower in the section, but also coarser and thicker conglomeratic beds. In contrast, the approximately 4,000 m thick Densuş-Ciula Formation covers a large area of the northwestern part of the Haţeg Basin.[40] Its gray-colored lower portion consists of repetitive sequences of volcanic and terrestrial strata, revealing a turbulent history punctuated by volcanic eruptions, while the upper portion is dominated by red conglomerates, sandstones, and mudstones.

The kinds of sediments in each of these Formations provide information about the environment in which they were deposited. Those of the Sânpetru Formation were laid down within rapidly flowing, braided river systems, probably deposited at the base of the newly exposed Retezat Mountains.[41] The Densuş-Ciula Formation appears to consist of strata deposited near or along the flanks of volcanoes.[42] The lower part of the Densuş-Ciula Formation is periodically interrupted by ashfall and mudflow deposits that contain volcanic debris, while the upper part lacks such beds, suggesting that this volcanic activity had decreased in the region.[43] According to Dan Grigorescu, the river systems of the latest Cretaceous within the Haţeg Basin flowed predominately northward, away from their sources in the Retezat Mountains.[44] The angularity of the grains and an abundance of feldspars and mica in the sandstones suggest that these sediments settled out in close proximity to their source. The conglomerates were deposited in rapidly flowing channels, while the sandstones and silty mudstones were laid down as river channels and bars, as well as overbank deposits during seasonal floods that burst the banks of the rivers and spread over the surrounding floodplain.

In the southern part of the Transylvanian Depression, the Upper Cretaceous Şard Formation is exposed along the southeastern margin of the Apuseni Mountains, between the village of Vurpăr and Pâclişa on the northern margin of the Mureş River, and near Sebeş.[45] Here it consists of a 2,500 m stack of red sandstone and mudstone beds that represent continental fluviolacustrine deposits. Moderately to well-developed paleosols, often associated with abundant bioturbation, developed on top of the abundant overbank deposits. Unlike the situation in the Haţeg Basin, the rivers flowed southward, away from their sources in the southern Apuseni Mountains. The Jibou Formation along the more northerly margin of the Transylvanian Depression is as much as 1,200 m thick.[46] Formed of coarse clastic rocks representing fluvial channels and red clay overbank deposits, it has only recently begun to yield much of a Late Cretaceous fossil record.

Based on studies of its Late Cretaceous vegetation, Transylvania probably had a tropical/subtropical climate, suggesting an annual mean temperature of approximately 22°C and annual precipitation of about 150 cm/year.[47] Occasionally lakes and ponds occupied the landscape, while soils developed on the floodplains. Analyzed by François Therrien of the Royal Tyrrell Museum of Palaeontology (Drumheller, Alberta,

Canada), these ancient soils—called paleosols—indicate that the Transylvanian climate, although generally subhumid and monsoonal, also had periods of aridity. Rainfall during these drier seasons was reduced to a range of 71 to 85 cm/year, and it lowered the mean annual temperature to 11.5°–11.7° C.[48] Evidence for these dry times come from the carbonate nodules—known from the Sânpetru, Densuș-Ciula, and Șard formations—that were formed by the precipitation of calcium carbonate as the moisture in the soil evaporated. These paleosols indicate that evaporation must have exceeded rainfall during part of the year. Bioturbation, in the form of fingerlike burrows and root traces a few centimeters long and several millimeters wide, riddles the red mudstones of the overbank deposits and even some sandstones. The burrows, probably feeding traces of annelid worms and insects, indicate that the floodplain and the other sediments were undergoing active turnover.

During the wet season, Transylvania was a mosaic of greenery, revealed by the relatively rich palynofloras and mega- and mesofloras from the Hațeg and Rusca Montană basins and the Apuseni Mountains. These floras—found in the alluvial fan and the volcanoclastic, fluviolacustrine, and coal-bearing deposits—consist of a mixture of horsetails (sphenophytes), ferns (pteridophytes), and flowering plants (angiosperms).[49] Dispersed conifer trees are present, while cycads, ginkgoes, and benettitaleans are rare or absent from these floras. Whether this paucity is biological (i.e., they did not live within the depositional basins) or taphonomic (i.e., they were present but their remains were not preserved) is unknown. Nevertheless, benettitaleans (cycadlike close relatives of angiosperms) and conifers, along with ferns and angiosperms, are abundantly known in the Bohemian Massif of Czechia from earlier times (Cenomanian; approximately 95 million years ago), and their presence in the latest Cretaceous of Romania, though speculative, might be anticipated.

We can reasonably assume that the shady river margins and moist marshy habitats in Transylvania were colonized by stands of horsetails, some of which grew to approximately 1 m high and were towered over by tropical ferns that probably stood up to 2 m high. Away from the rivers, in patches of sunlight, horsetails and ferns flourished in the loose and gravely, seasonally wet soils. The better-drained habitats farther from the rivers were dotted with a heterogeneous mix of ground cover, shrubs, and

Figure 4.7. Probable elements of the Late Cretaceous flora of Transylvania. Arborescents: palm (Arecaceae), sycamore (Platanaceae), walnut (Juglandaceae), and laurel (Lauraceae) (*above, from left to right*). Ground cover: horsetail (Equisetaceae), fern (Polypodiaceae), and clubmoss (Lycopodiaceae) (*middle, from left to right*). Examples of leaves from dicot (*bottom left two*) and monocot (*bottom right two*) plants known from the Rusca Montană flora. Scale of leaves = 5 cm; the figures of the arborescents and the ground cover are not to scale. (After Petrescu and Duşa 1982)

stands of trees (figure 4.7). What the herbaceous ground cover might have resembled is unknown, but it probably consisted of small ferns, sphenophytes, and club mosses (lycophytes). Shrubbier understory vegetation was composed principally of angiosperms and low tree ferns. The canopy of the open woodlands probably extended to a height of 20–30 m and was formed of filicalean ferns (Gleicheniaceae, Polypodiaceae), broadleafed conifers (?Cheirolepidiaceae), and woody angiosperms, including palms (Arecaceae), screw pines (Pandanaceae), walnuts (Juglandaceae), and laurels (Lauraceae). However, it was the sycamores (Platanaceae) that appear to exhibit the greatest diversity and biomass of these Transylvanian forests.[50]

The only invertebrates known from the Late Cretaceous of the Transylvanian region are small land snails, freshwater unionid clams, and ostracodes.[51] Besides their biostratigraphic significance, these mollusks provide additional insight about the paleoecological relationships of the area. Like their present-day relatives, the Hațeg snails probably required constant moisture, food, and shelter, most likely living beneath rocks and the broadleaf litter of the forest floor. There, they crept about, each on its broad, flat, flexible foot, feeding upon detritus and each other or scavenging on the dead and dying. As for the unionids, these freshwater bivalves were burrowers, reworking the sediments on the bottoms of streams and rivers in order to feed on small food particles. As such, they formed an important part of the Hațeg freshwater ecosystem.

Other invertebrates almost certainly were present in the Late Cretaceous of Romania, but the fossil record for this part of the fauna is virtually silent. This applies especially to insects, for which the Transylvanian record is thus far absent (except for a dermestic beetle boring into a *Magyarosaurus osteoderm*[52]) and, indeed, is rather scanty throughout the remainder of Europe during this interval. This limited evidence of insect activity includes leaf damage, eggs, and body parts of butterflies, dragonflies, beetles, and caddisflies from the earliest Late Cretaceous of the Bohemian Massif (Czechia) and the Paris and Aquitanian basins of northwestern France.[53] However, we do know from the records elsewhere, and from earlier times, that virtually all major groups of extant insects had evolved by the Maastrichtian. Based on this information, it's reasonable to assume that the underbrush and leaf litter of Transylvania must have abounded with insect life. Termites, ants, and cockroaches tunneled through the detritus. Grasshoppers, crickets, and katydids hunted and chirped, while cicadas sang their whining all-day, all-night song. Beetles, bugs, mantises, and aphids held court in the trees and shrubs. In the air, bees, wasps, flies, and butterflies buzzed and flapped, while dragonflies silently hovered above it all.

Here we have presented a sketch of what the environmental setting of

Transylvania might have been like some 70 million years ago. Some of the information is readily available from the rock and fossil record, but other aspects remain highly speculative. On this basis, do we now think that the paleogeography, paleoenvironment, and biota is well understood? Not yet we don't.

HAȚEG ISLAND?

As we have noted, Nopcsa and several recent paleogeographers of the Mesozoic of Europe depicted western Romania as a reasonably large island, often within an archipelago whose activity is due to movements of small tectonic plates associated with the conjoined motion of Africa and Apulia. However, things are not necessarily so straightforward; other studies indicate that this habitat reconstruction is far too simple.[54] We cannot overemphasize how extremely difficult it is to reconstruct now-obliterated habitats, especially those terrestrial ephemera created and disposed of by tectonic dynamism during the evolution of the Cretaceous Neotethyan Ocean. Despite his strong belief that the Hațeg area was part of an island, Nopcsa also recognized that there was a paucity of geologic evidence to suggest that this was true—no trace could be found of beach or other tidally influenced deposits indicating the periphery of such an island. Abel also criticized Nopcsa's reconstruction of the Hațeg region as an island, questioning whether the area that supported the fauna was actually small enough to be considered an island or, instead, was sufficiently large in this regard to be considered a mainland.

Just how do we identify islands in deep time? Unlike the present or recent past, where islands can be quite easily seen or are uncontroversial in their reconstruction, we need to be more careful with our prehistoric interpretations. Nopcsa's strategy was to enlist both geology and biology, just about the only observations we still have available to us. From geology, we could look for sedimentological evidence of beaches or shorelines (which Nopcsa—and others since—have been unable to find). We could also look to the depositional context, in particular the thickness of the stack of terrestrial sediments, for an estimate of the size of the rapidly subsiding depositional basin. Given that the thicknesses of the terrestrial rocks observed in the Hațeg Basin and the Transylvanian Depression range from 2,500 to 10,000 m, this large accumulation of sediment means that the Hațeg region cannot be reconstructed as a small island,

and it needed to have been a sufficiently large emergent area to be able to accommodate the tectonic activities that clearly occurred there. As we have seen in the earlier portions of this chapter, another source of data comes from paleogeographic reconstructions. Current paleogeographic maps indicate a sizeable emergent area in this northern Peri-Tethyan realm, corresponding to the relatively small Hațeg Basin and much larger Transylvanian Depression. However, from what is now eastern Iran to Italy, the entire area was a huge swath of active mountain building (Plate VI).[55] It is therefore difficult (if not impossible) to assess, amid all of this flux, many of the geographic details necessary to have made Transylvania a sufficiently sizeable and remote island.

Fortunately, the best indicators of islands seem to be biological.[56] The nature of their isolation from more continental faunas will produce different effects on island faunas. First, it is highly likely that island faunas will be less diverse than those on the nearby mainlands from which they colonized. Second, these faunas are very likely to be unbalanced, with major clades known from the mainlands absent from the islands. Third, depending on how long ago the islands were colonized, their faunas may resemble the more primitive nature of earlier faunas elsewhere. Fourth, for those taxa present on islands, there is probably a high degree of endemism. Finally, for the individuals living on islands, changes in body size are common.

The issue of body size will be considered in some detail in chapter 5. Here, we will explore diversity, balance, primitiveness, and endemism in Transylvania, based on information about the membership of other faunas. We begin by comparing the Transylvanian fauna with other assemblages from Europe (figure 4.8). These include several sites in Sweden, England, Bulgaria, and Czechia; numerous localities in southern France, Spain, and Portugal; the assemblage from Muthmannsdorf in Austria; several nearshore marine sites around Maastricht in the Netherlands and adjacent parts of Belgium; a poorly known assemblage from Crimea in Ukraine; and new finds made in the late 1990s and early 2000s from Italy, Germany, Slovenia, and Hungary.[57]

Paleontologists working at sites in southern France have recovered some of the most important terrestrial and nearshore marine vertebrate faunas from the Late Cretaceous of Europe.[58] From Provence in the east to the northern foothills of the Pyrenees in the west, a wealth

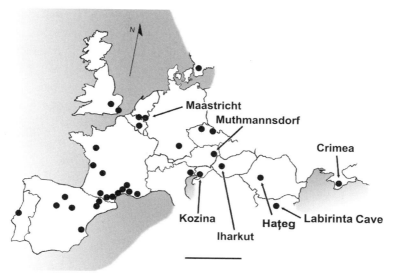

Figure 4.8. A map of modern Europe, indicating known dinosaur-bearing localities from the Late Cretaceous. Scale = 600 km

of dinosaurs, turtles, crocodilians, pterosaurs, bony fish, frogs, lizards, and mammals have been collected since the earliest discoveries in the 1700s.[59] There are theropods from southern France, quite similar to the dinosaurs of the Transylvanian fauna. Like their Transylvanian relatives, most of these French specimens are poorly understood as yet. They include small forms, as well as the 7–10 m long *Tarascosaurus* and some tantalizing remains of a large bird (neither of which is yet known from Transylvania). A titanosaurian sauropod is present—*Ampelosaurus*—as are the ornithopods *Rhabdodon*, some newly discovered hadrosaurid material referred to *Pararhabdodon* (see below), and the nodosaurid *Struthiosaurus*. There are also plenty of turtles, fish, and crocodilians, gigantic pterosaurs, and eggs displaying a megaloolithid architecture.[60]

Numerous terrestrial faunas have also been found across the Iberian Peninsula. Those from Portugal are less well known, but they include, among the dinosaurs, a variety of small theropods, titanosaurian sauropods, nodosaurids, and ornithopods, as well as crocodilians, turtles, mammals, amphibians, and fish.[61] Spanish localities have yielded a variety of thus-far poorly understood small theropods, titanosaurian sauropods (including one named *Lirainosaurus*), the nodosaurid *Struthiosau-*

Living on the Edge 107

rus, at least two kinds of ornithopods (*Rhabdodon* and *Pararhabdodon*), dinosaur eggs, pterosaurs, and a rich array of crocodilians, lizards, snakes, turtles, placental mammals, amphibians, bony fish, and sharks.[62]

We have been introduced already to the Gosau fauna of Muthmannsdorf, Austria (chapter 2).[63] The dinosaurs from this area consist of indeterminate theropods, an indeterminate rhabdodontid (formerly known as *Mochlodon suessi*),[64] and the nodosaurid *Struthiosaurus*, while the rest of the fauna includes pterosaurs, crocodilians, a choristoderan (a small crocodilianlike fish-eater), lizards, and turtles.

Nearshore marine deposits from the Maastricht region of the southernmost Netherlands and adjacent Belgium have produced a rich marine fauna, but very few dinosaurs thus far.[65] A hadrosaurid known as *Orthomerus dolloi* and a theropod called *Betasuchus bredai* are known only from isolated and often fragmentary bones and teeth. These remains may represent cast-off body parts from bloated carcasses that drifted out to sea, for otherwise the vertebrate fauna consists of marine crocodilians, plesiosaurs, mosasaurs, turtles, bony fish, and sharks.

Of the few remaining European faunas, the earliest known, described in 1945 by A. N. Riabinin from St. Petersburg, is from Crimea, the portion of Ukraine projecting into the Black Sea.[66] Several postcranial skeletal elements belong to a hadrosaurid (originally called *Orthomerus weberi*, but this is probably not diagnostic). More recently, Peter Wellnhofer, a Bavarian paleontologist and one of the world's leading specialists on Pterosauria, described a tantalizing, yet frustrating hadrosaurid femur from Maastrichtian marine limestones in Bavaria, Germany.[67] At the opposite end of the scale of preservation, some excellent hadrosaurid material, including most of the skeleton and skull, occurs in Santonian deposits in northeastern Italy; it has been named *Tethyshadros insularis*.[68] Finally, discoveries made over the past ten years in several areas in eastern and northern Europe—Kozina, Slovenia; Iharkút, Hungary; Labirinta Cave, Bulgaria; Mezholezy, Czechia; and the Belgorod and Volgorod regions of Russia—have disclosed additional dinosaur records from the Late Cretaceous. Thus far the dinosaur fauna from Slovenia, which is roughly contemporary with those of southern France, includes hadrosaurids and other ornithopods, small theropods, and two different crocodilians,[69] while that from Hungary, dating from as early as late

Santonian, indicates the presence of a neoceratopsian (*Ajkaceratops kozmai*), a nodosaurid ankylosaur (*Hungarosaurus tormai*), a rhabdodontid ornithopod, theropods (including an enantiornithine bird), an azhdarchid pterosaur (*Bakonydraco galaczi*), several kinds of crocodilians (including *Iharkutosuchus makadi*), turtles, amphibians, and fish.[70] The Bulgarian occurrence is a disarticulated, incomplete hadrosauroid hind limb and a caudal vertebra that come from Labirinta Cave, southwest of the town of Cherven Bryag, in the northwestern part of the country.[71] From Czechia comes a fragmentary yet tantalizing medium-sized femur that is thought to come from an ornithopod dinosaur. It was collected from shallow marine sediments (late Cenomanian in age) near the village of Mezholezy, between the towns of Kutná Hora and Čáslav. Currently, the Russian material from Belgorod consists of a tooth and a vertebra of what appears to be a close hadrosaurid relative, perhaps similar to *Telmatosaurus*, but of Albian–Cenomanian age.[72] Finally, the Volgorod locality, Maastrichtian in age, has yielded two dinosaurs (an indeterminate hesperornithid and a basal theropod) and a few turtles.[73]

How do these European faunas compare with each other and with others globally, in terms of diversity, balance, primitiveness, and endemism? It can be argued that the Transylvanian fauna is not much different from those elsewhere in Europe at the end of the Late Cretaceous—it has approximately the same within-clade diversity of creatures as France, Spain, or Austria: one or two ornithopods, a titanosaur or two, small and rare theropods, maybe a nodosaurid, pterosaurs, eusuchian crocodylomorphs, lizards, and amphibians. From this quick census, we would be hard-pressed to see many differences in diversity among the European regions.

In this way, the European faunas appear to be reasonably ecologically balanced when compared with one another; that is, the makeup of the trophic organization of each region is roughly equivalent. For example, there are a variety of primary consumers in each fauna. Which group is at the top of the food chain (small theropods, pterosaurs, or eusuchian crocodilians) is somewhat questionable, but all are well represented from region to region. None of these faunas seems out of place with the others in terms of relative primitiveness. For all of these reasons, none of the European faunas appear to be loaded with endemics.

That said, we've been looking at the wrong level for our comparisons. Major clades are expected to be widely distributed, on mainlands and on islands. When looked at from a species-level perspective, however, the pattern of similarity across the region changes. *Telmatosaurus transsylvanicus* is known only from Romania, while *Pararhabdodon isonensis* comes just from Spain and France. *Ampelosaurus atacis* has been found only in France, while *Magyarosaurus dacus* hails solely from Romania. The three species of the nodosaurid *Struthiosaurus* (*S. austriacus*, *S. transylvanicus*, and *S. languedocensis*) are each found in different parts of Europe. At these lower taxonomic levels, each European fauna is completely endemic to its particular region.

Global comparisons of diversity, balance, and primitiveness are perhaps the most telling. At the global level, mainland faunas often have twice or three times (and more) the within-clade species diversity that is found in the different regions of Europe.[74] Then there's the issue of faunal balance, in particular regarding the top predator. If the European faunas of the Late Cretaceous differ in terms of the makeup at the top of the food chain, then they are not just microcosms of continental conditions elsewhere in the world. When compared globally, the most obvious difference between these European assemblages and those elsewhere in the world is who is represented in this top rank. The absence of a large theropod in Transylvania and in the other European faunas differs from the more continental situation during the Late Cretaceous. In western North America, we have *Albertosaurus*, *Daspletosaurus*, and *Tyrannosaurus*; central and eastern Asia has its *Tarbosaurus* and *Alectrosaurus*; and in South America there are *Abelisaurus* and *Giganotosaurus*.[75] The lack of a large theropod in the European assemblages could be an artifact of collecting; none of the European localities can match the availability of fossils in the badlands and deserts of these other, larger regions. Nevertheless, if this lack is not an artifact, then the European faunas do depart from the more continental conditions of North and South America and Asia, making them unbalanced, at least in terms of who the top predator was. Further faunal absences from Europe consist of a host of major theropod groups, including tyrannosauroids, ornithomimosaurs, oviraptorosaurs, and therizinosauroideans. Another factor in this imbalance is the absence of diplodocoidean sauropods, ankylosaurids, pachycephalosaurs, and ceratopsians in Transylvania. Without many of these taxa, the European faunas tend to look rather primitive, more closely resembling those from the Early rather than the Late Cretaceous.

INSULARITY IS IT

The Transylvanian region seems to have been an island or, at the least, it behaves like one biologically. Even so, its areal extent is nearly unconstrained. It assuredly was large enough to form basins with very thick terrestrial depositions. It was also big enough to support numerous interacting tetrapod species. Could it have been large enough or close enough to contemporary mainlands to operate as an outpost, part of a tectonically active region where sizeable terrestrial habitats were contiguous with continental landmasses, but on an off-and-on basis?[76] At the moment, it's difficult to say. And maybe it's not really relevant to determine whether the Transylvanian region was an island or an outpost. Perhaps what is most important here is the degree of insularity or isolation of the Transylvanian region and its fauna from those elsewhere in the world.

Insularity need not be limited just to those landforms we call islands. Rather, it is used as a qualitative measure of isolation in present-day ecosystems. MacArthur and Wilson stated this in their opus on theoretical ecology and island biogeography:

> Insularity is moreover a universal feature of biogeography. Many of the principles graphically displayed in the Galápagos Islands and other remote archipelagos apply in lesser or greater degree to all natural habitats. Consider, for example, the insular nature of streams, caves, gallery forest, tide pools, taiga as it breaks up into tundra, and tundra as it breaks up into taiga.[77]

In the case of our Transylvanian investigation, what matters is the insularity of the region, rather than the name of the landform we give to where it existed. Islands surrounded by water, remote lakes surrounded by land, clumps of woodland in a prairie, oases in the desert, mountain peaks, geologically ephemeral outposts in a tectonically active assemblage of microcontinental plates—all are characterized by being isolated from each other by a formidable barrier.[78] However we reconstruct the paleogeography of Transylvania, it was certainly an insular place, by virtue of tectonics, geography, and ecology. The Transylvanian dinosaurs made their living on the edge, thanks to their insularity from the much larger continental faunas of Africa, Asia, and North America. Now it is a matter of understanding how and when this remarkable fauna may have arisen. To do so, we will take up the issues of phylogeny, heterochrony, and historical biogeography in chapters 6, 7, and 8.

PLATE I. *Top*: Our Transylvanian camp on the Sibişel River, 1994. *Bottom*: Our field crew, 1995

PLATE II. *Top*: Franz Baron Nopcsa (1877–1933). *Bottom*: The Nopcsa castle at Săcel, Romania

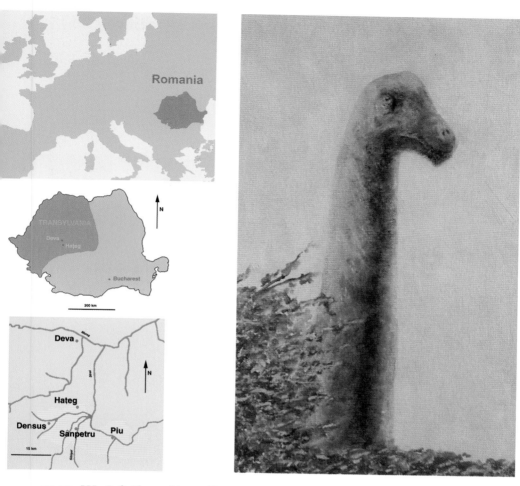

PLATE III. *Left*: The positions of Romania in Europe (*top*); Transylvania in Romania (*middle*); and the Haţeg Basin in Transylvania (*bottom*). *Right*: A restoration of the titanosaur sauropod *Magyarosaurus dacus*

PLATE IV. *Top*: A restoration of the "Romanian raptor." *Bottom*: A restoration of the nodosaurid ankylosaur *Struthiosaurus transylvanicus*

PLATE V. *Top*: A restoration of the ornithopod *Zalmoxes robustus* (after "Walking with Dinosaurs," 1999. BBC Television). *Bottom*: A restoration of the ornithopod *Telmatosaurus transsylvanicus*

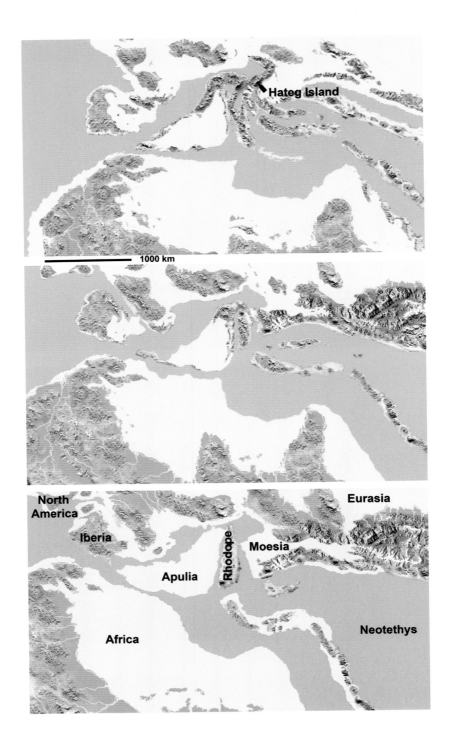

PLATE VII. *Top*: The Retezat Mountains. *Bottom left and right*: An outcrop of the Sânpetru Formation along the Sibişel River

PLATE VI. *Opposite.* Paleogeographic maps of the Tethyan realm during the Early Cretaceous (*bottom*), mid-Cretaceous (*middle*), and Late Cretaceous (*top*). North America, Iberia, Africa, Apulia, Rhodope, Moesia, Neotethys, and Eurasia are indicated on the Early Cretaceous map. Haţeg Island is indicated on the Late Cretaceous map (after R. Blakey, www2.nau.edu/rc67/).

PLATE VIII. Outcrops in the Haţeg Basin:
Lower Quarry, Sânpetru Formation (*top left*);
Tuştea, Densuş-Ciula Formation (*top right*);
and Pui, Sânpetru Formation (*bottom*)

CHAPTER 5

Little Giants and Big Dwarfs

The earliest recognition of fossil bones, whether they were dinosaurian, mammalian, or some other vertebrate, cannot be recounted, but it is clear that they were given serious attention at least since antiquity. Seized on as being the remains of giants, so began the link between fossils and what might be called gigantology, a fascination with large size that continues to this day.[1]

It was the bones of large fossil mammals, especially of extinct elephants, that originally attracted attention. Adrienne Mayor, in her tour de force treatment of the meaning of fossils in antiquity, noted many instances of massive bones discovered in such places as the islands of Sicily and Capri that were thought to be the remains of a race of giants.[2]

Discoveries and interpretations such as these continued to be made throughout the Middle Ages and the Renaissance—indeed, up until the end of the seventeenth century—whereby large fossil bones were generally thought to be the remains of human giants, dragons, or mythical monsters hearkening back to Greek mythology. In the skull of a Pleistocene dwarf elephant discovered in a Sicilian cave near Trapani in the fourteenth century, Giovanni Boccaccio (1313–1375), author of the *Decameron*, saw the face of the cyclops Polyphemus.[3]

As Christianity took hold across Europe, scripture ("There were giants in the earth in those days," Genesis 6:4) and the existence of ancient behemoths became closely tied. Consequently, the practice of attributing fossil bones to giants—often as saints and other biblical personages—led to these remains being kept in churches, especially in Europe. Other interpretations fashionable at that time included ascribing animal fossils to real people. Thus, according to legend, the remains of Theutobochus (King of the Teutons, Cimbri, and Ambrones) were recovered at Langdon, France, in 1613. The giant Theutobochus was the basis for considerable controversy, in which opposing physicians and surgeons attacked each other over the authenticity and determination of these remains. In the end, these bones proved to be those of a Miocene elephant called *Deinotherium giganteum*.[4]

A more modern interpretation of fossil bones began to emerge with the rise of interest in comparative anatomy. During the seventeenth century, Robert Plot (1640–1696), curator of Oxford University's Ashmolean Museum, was busy producing the first illustrated book of fossils from England.[5] Among the plates was a figure of the lowermost part of a thighbone that looked very similar to that of humans, but was much greater in size (figure 5.1). Plot concluded that it "must have belonged to some greater *animal* than an *Ox* or *Horse*; and if so in all probability it must have been the *Bone* of some *Elephant*, brought hither during the Government of the Romans in Britain."[6] Another century passed, and Richard Brookes (dates unknown), also a natural historian from England, copied Plot's figure into a compendium of natural history.[7] In a caption to this illustration, Brookes now applied a name—*Scrotum humanum*. Although his designation was in apparent recognition of its general shape, Brookes was not persuaded that the fossil was an actual fossilized gigantic scrotum (although one French philosopher, Jean-Baptiste Robinet [1735–1820], apparently was convinced[8]). Instead, Brookes considered it, as Plot had before him, a portion of the thigh bone of a large animal. Although the original specimen is lost, we now know that this fragment was the lower end of a femur from the Middle Jurassic of Oxfordshire and that it must have belonged to some sort of large theropod dinosaur.[9]

In the seventeenth century, dinosaurs, and indeed deep time itself, had yet to be discovered; it took three people from England, the help of the great French comparative anatomist Georges Cuvier, and the first half of the nineteenth century to finally recognize them.[10] William Buckland (figure 5.2, left), Reader of Mineralogy, Reader of Geology, and then

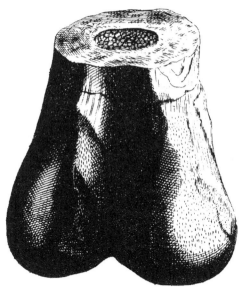

Figure 5.1. The first "named" dinosaur, *Scrotum humanum*, in actuality the lower end of a *Megalosaurus* femur. (Original plate from Brookes 1763)

Canon of Christ Church College at Oxford University, sought to fuse geology and paleontology with the traditional Christian teachings of Noah's flood and a divine Creator. A legendary eccentric, Buckland is also credited with the first scientific description of a dinosaur, a theropod that he named *Megalosaurus*, based on a fragment of a lower jaw with teeth and some postcranial bones from the Middle Jurassic Stonesfield Slate of Oxfordshire.[11] Buckland's contemporary, Gideon Mantell (figure 5.2, right), was trained as a physician, a career marked by considerable failure. It is Mantell's other life pursuit, research on the paleontological riches of the English Weald, for which he is now remembered. As one of the world's original dinosaur hunters, he energetically collected, swapped, and purchased teeth, isolated bones, and portions of a skeleton, from which he gave us the first-discovered herbivorous dinosaur, *Iguanodon*, from the Early Cretaceous of southeastern England.[12]

Together, Buckland and Mantell recognized the unique nature of their material: the bones and teeth of reptiles whose great size eclipsed any reptile and most mammals known up to that time. Mantell, in particular,

Figure 5.2. William Buckland (1784-1856; *left*) and Gideon Algernon Mantell (1790-1852; *right*)

had the vision to impart animal form and habits to these fossil bones, as well as the intellectual courage to perceive that these were animals of the remote past, not relics of Noah's flood. Deep time was about to be discovered.

The actual recognition of Dinosauria, including *Megalosaurus*, *Iguanodon*, and an armored dinosaur known as *Hylaeosaurus*, was left in the hands of Sir Richard Owen.[13] A comparative anatomist at the Royal College of Surgeons in London, Owen built on his personal insights and those of Mantell and Buckland and, with a flash of insight and no little political guile, hatched Dinosauria in April 1842.[14] "The combination of such characters [in particular, a sacrum composed of more than two fused vertebrae] all manifested by creatures far surpassing in size the largest of existing reptiles, will, it is presumed, be deemed sufficient ground for establishing a distinct tribe or suborder of saurian reptiles, for which I would propose the name of Dinosauria."[15] This new group of reptiles called Dinosauria—the "fearfully great lizards"—was thought to consist of highly advanced terrestrial quadrupeds with the unmistakable character of enormous size.

From Owen's day onward, the quintessential dinosaur, at least in the popular imagination, has been gigantic. These creatures were and still are the behemoths that pack the public into museum exhibits.[16] With each thud of their feet, they have rumbled their way into books of all sorts, onto postage stamps and collecting cards, into children's games (including the computer variety), and, of course, onto the big screen, from *Gertie the Dinosaur* (1912) through *Jurassic Park III* (2001).[17]

Dinosaur size records are set, only to be broken. Notable among the carnivores, the perennial favorite, *Tyrannosaurus* (or *T. rex* to its friends), a 12 m long predator from the Late Cretaceous of North America that may have tipped the scales at 7 metric tonnes, now has plenty of company among the great Mesozoic slayers.[18] *Carcharodontosaurus*, a theropod from the middle Cretaceous of northern Africa, has been estimated to be as long as *Tyrannosaurus*, but it appears to have been slightly more massive,[19] whereas *Giganotosaurus* from the Early Cretaceous of Argentina is thought by some to measure over 14 m and weigh up to 8 metric tonnes.[20] Yet among the truly supergigantic, there's nothing like the sauropods, which have long been known to be the largest of all terrestrial animals ever to have lived. For many years, the prize went to *Diplodocus*, reaching a length of 27 m,[21] but this record has seen its challenges. In North America, there is *Seismosaurus*, an as-yet poorly known sauropod from the Late Jurassic of the southwestern United States, whose length has been variously estimated at upwards of 50 m.[22] Not to be outdone, the Southern Hemisphere has offered up *Argentinosaurus*. Discovered in 1989, named in 1993, but with other aspects not yet fully published, this behemoth has also been estimated to be as much as 50 m long.[23] These are clearly humbling animals.

The evolution of dinosaurian gigantism is but one example of the general trend of animals becoming larger over time. Other well-known examples include horses, ratites (ostriches and their kin), and ammonites (long-extinct relatives to today's pearly nautilus, which grew to diameters of about 2 m). Such evolutionary trends in increasing body size—commonly referred to as Cope's Rule, after E. D. Cope's recognition that evolution often results in phylogenetic size increases[24]—are thought to be due to an extension of growth tendencies already present in ancestral ontogenies. These extensions are known as peramorphs ("beyond shapes").[25] Peramorphosis implies that descendants reach new end stages (for example, a larger body size) by passing along this course of ancestral ontogeny, but then, instead of standing still, advance still further.

TRANSYLVANIA: BUCKING THE GIGANTISM TREND

The production of peramorphs through an extension of growth has its opposite manifestation in the production of paedomorphs ("baby shapes") through arrested growth.[26] Said another way, paedomorphosis is the retention of ancestral juvenile characters into the adulthood of its descendants. Referred to by Ken McNamara (paleontologist and evolutionary biologist at the Western Australian Museum in Perth) as the Peter Pan syndrome, two modern organisms stand out as the most celebrated examples of arrested growth.[27] The first is the axolotl (figure 5.3), otherwise known as the Mexican salamander *Ambystoma mexicanum*, which retains its larval characteristics into adulthood and thereby is able to reproduce while remaining in the form of an aquatic larva.[28] The other modern epitome of paedomorphosis, as we shall see a little later, is ourselves—*Homo sapiens*. Furthermore, we encounter paedomorphs among dogs, horses, mice (including one named Mickey), finches, ratites, fish, insects, snails, and plants. They are even known from the fossil record, including the dwarfed dinosaurs from Transylvania.[29]

In chapter 2, we let Franz Baron Nopcsa introduce, for the first time, the notion of small dinosaurs from Transylvania in the context of his hypothesis that in the Cretaceous this region was an island. In his first synthesis of this subject, Nopcsa surveyed the Transylvanian fauna,[30] emphasizing in particular the disparity in body size of its members: "While the turtles, crocodilians, and similar animals of the Late Cretaceous reached their normal size, the dinosaurs almost always remain below their normal size."[31] He observed that most of the Transylvanian dinosaurs hardly reached 4 m in length, and the largest (what was to become *Magyarosaurus dacus*) was a puny 6 m long, compared with a more representative 15–20 m for other sauropods. After this 1915 paper, Nopcsa's fascination with body size shifted to dinosaurian giants, but in both cases he equated the evolution of both dwarfism and gigantism to changes in the endocrine system, such as in the size and function of the pituitary gland.[32]

Nopcsa viewed Transylvanian dwarfing as the body-size consequences of island evolution. Thus these dinosaurs joined the ranks of other insular organisms, both living and extinct, that have evolved into smaller and larger versions of their continental relatives: dwarfed crocodilians, giant

Figure 5.3. The axolotl (*Ambystoma mexicanum*), a living salamander that reproduces as an aquatic form; in this way, it is a modern exemplar of paedomorphosis. (After Lamar 1997)

moas, downsized island deer, enormous Galápagos turtles, dwarfed donkeys, huge Komodo lizards, small Bali tigers, pygmy elephants, and dwarfed hippos.[33] After 1914, Nopcsa's efforts to define "small" in dinosaurs ceased, and he simply concluded that the Transylvanian dinosaurs were "less than the usual colossal forms" that tended to be found elsewhere. Whereas such a statement is certainly true, it belies a more complex set of questions about the acquisition of small body size. Yes, these dinosaurs were smaller, but were they, and they alone, miniaturized from a larger ancestor?

Heterochrony: Timing (and Rate) Is Everything

Why is it that all babies are born looking like Winston Churchill, and vice versa? To put it another way, why do the faces of both young and old humans alike tend to look more like the faces of short-faced chimpanzee and gorilla youngsters than these latter two juveniles do to their respective long-faced parents (figure 5.4)?[34] These similarities and differences in proportions have been attributed to changes in different aspects of facial growth and, when given an evolutionary context, are described in terms of heterochrony. We have already seen heterochrony in action through the gigantic (peramorphic) tendencies of dinosaurs and the paedomorphosis of the axolotl. In the case of humans, gorillas, and chimps, it is the retardation of growth in the jaw region that gives us that youthful look. So to the degree that the pattern of growth in chimps and gorillas is considered the primitive condition for the immediate clade of primates that includes these two groups and us, then we (or rather our faces) represent the paedomorphic retention of juvenile features into adulthood.

In their book of the same name, Michael McKinney and Ken McNamara define heterochrony (meaning "different time") as the "change in timing or rate of developmental events, relative to the same events in the ancestor."[35] In other words, heterochrony seeks to link development with evolution. In Darwin's world, where external forces held sway, there was little attempt to link ontogeny with phylogeny through his view of natural selection. Instead, it was principally the German scientific community, notably the nineteenth-century school of *Naturphilosophie*, that attempted to link developmental sequences with similar patterns found in evolutionary history.[36] Ernst Haeckel (1834–1919), in his attempt to reconcile Darwin's evolutionary mantra of "descent with modification" with emerging issues in developmental biology, coined his own slogan, "ontogeny recapitulates phylogeny," now known as Haeckel's Biogenic Law.[37]

The most fundamental reconsiderations of how development interconnects with phylogeny to produce heterochrony have been through Stephen J. Gould's now-classic book, *Ontogeny and Phylogeny*,[38] and a subsequent paper cowritten by him and an impressive cadre of evolutionary and developmental biologists: Pere Alberch, George Oster, and David Wake.[39] Both publications stress how the features of organisms change with size during both ontogeny and phylogeny, and the paper sets forth the schema of heterochronic terminology to express these changes. The concept of heterochrony has seen an incredible growth in evolutionary biology in recent years.[40] It is now used to explain, with great success, the differences in skull forms of domesticated dogs, the evolution of fossil

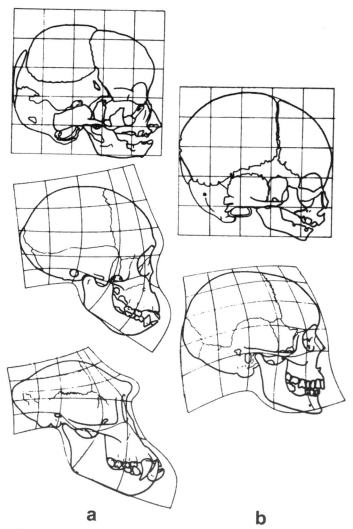

Figure 5.4. Neoteny and the human face. The left column (a) of three skulls represents the ontogenetic changes in the facial skeleton of a chimpanzee, while the right column (b) of two skulls represents a newborn and an adult human. The superimposed grids indicate how much change occurs in the skulls during ontogeny. Note the similarity of the faces of a newborn chimpanzee and a newborn human, the similarity of newborn and adult humans, and the dissimilarity between an adult human and an adult chimpanzee. (After Starck and Kummer 1962)

sea urchins, the outrageously large antlers in the so-called Irish elk (actually a giant deer), and sexual dimorphism in the forked fungus beetle.[41] Each of these examples of heterochrony, and many more, describes how ontogenetic changes form the basis for phylogenetic transformations.

So what is this thing we call heterochrony? We began talking about the subject by introducing paedomorphs and peramorphs, the two ontogenetic patterns that are linked between ancestors and descendants. A paedomorph is an ontogenetically less well-developed descendant than its ancestor. That is, reduced or arrested development will produce a paedomorph. In contrast, if its development goes beyond that of its ancestor, a descendant is considered to be a peramorph. Peramorphs are produced by increased development. A sequence of successive paedomorphs is known as a paedomorphocline. Likewise, a continued trend of peramorphs is called a peramorphocline. Each of these states—paedomorphosis and peramorphosis—is the product of three growth processes: (1) the timing of trait formation, (2) the cessation of its development, and (3) the rate at which it is acquired (figure 5.5).

A change in the time that the growth of an organ or structure starts development, relative to others, results in either predisplacement (if the change is earlier) or postdisplacement (a later change). Predisplacement is seen in the nasal horns of the large, early Tertiary mammals known as titanotheres, in which earlier development results in peramorphosis.[42] In contrast, postdisplacement involves the delayed onset of growth as, for example, is seen in the timing of the cellular differentiation of fins and limbs in vertebrates.[43] In this way, later starters produce the paedomorphic condition.

When the development of a structure ceases also has an effect on the final body form of an organism. Earlier cessation/offset in a descendant, rather than in its ancestor, produces what is known as progenesis. Progenesis is thought to have altered the ancestral developmental program of arthropods and reduced the number of segments and limbs in insects, the paedomorphic descendants of their common ancestor with the multisegmented, multilegged centipedes and millipedes.[44] Hypermorphosis produced by delayed offset—the extension of growth beyond that of an ancestor—has peramorphically produced large dogs such as Irish wolfhounds, Great Danes, and St. Bernards.[45]

Changes in the rate of growth can produce either acceleration, when

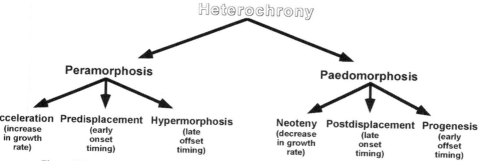

Figure 5.5. Heterochrony, its two manifestations (peramorphosis and paedomorphosis), and the six processes that produce them. (After McKinney and McNamara 1991)

growth rate is increased, or neoteny, when growth rate is retarded. Neoteny—that is, less shape change and a decrease in complexity—is thought to be behind why adult human faces look much like those of youngsters: the rate of growth of the face is decelerated, compared with that seen in ancestral ontogenies, to produce the paedomorphic human facial profile.[46] Acceleration, by contrast, can be seen in the growth of the bones in the digits of the hand in bats, acting to produce the peramorphic scaffolding of their special kind of airfoil for flight.[47]

As we have briefly introduced the concepts here, when descendants go beyond the size and shape of their ancestors, we have a pattern known as peramorphosis. And when descendants fail to achieve the size and shape of their ancestors, due to arrested growth, we have paedomorphosis. Although peramorphs and paedomorphs can be more-or-less easily recognized by allometric comparisons of size and shape, the processes that produce them (e.g., neoteny, progenesis, postdisplacement, etc.) are more difficult to discern, because they require a measure of actual ontogenetic timing. For living organisms, there is no barrier to establishing when various life-history events—the timing of protein expression, cell differentiation and interaction, hormone secretion, sexual maturity, and so on—take place (although this information is rarely available). The question is, can we do this with fossils?

We are fortunate to be living in a time when the insides of fossils are as important—and are nearly as easily seen—as their outsides are. Bone in thin-section is beginning to reveal how dinosaurs lived, reproduced, metabolized, and grew in ways for which, a generation ago, we had hardly a clue. For our purposes here, bone in thin-section appears to be able to allow us to calibrate anatomical changes in real time. This burgeoning research is called skeletochronology.[48]

How does it work? First, we know that bone growth can be episodic, and this ebb and flow is laid down in the bone itself, much like tree rings.[49] These lines, found in both extant and extinct vertebrates, reflect disruptions in growth: periods of rapid growth followed by slowdowns or cessation. More simply put, growth lines—what scientists call lines of arrested growth (LAGs)—appear as thin, circumferential, avascular regions when bone is seen in thin-section. LAGs are thought to be annual, representing yearly histories of seasonality and ecological stress.

Provided that they are annual, LAG counts give an estimate of the longevity of a particular individual. When these are combined from individuals of different sizes within a single species—say, *Tyrannosaurus rex*—such longevity estimates, coupled with size data or mass estimates, can be used to make the classically sigmoidal age-versus-size curves. From such studies, we know that at least some of the extinct dinosaurs (e.g., ornithopods, theropods, and sauropods) grew quite rapidly, perhaps as much as is seen in modern birds.[50] Yet how did these rates vary phylogenetically among close relatives? By reconstructing and comparing the growth curves of closely related forms, it is possible to bring skeletochronologies to bear on real-time heterochronic problems. This has indeed been done for *T. rex* and other tyrannosaurids, *Janenschia* and other sauropods, the prosauropod *Plateosaurus*, the ceratopsian *Psittacosaurus*, and, perhaps most relevant here, a new species of dwarfed sauropod from the Late Jurassic of Germany.[51]

Preliminary thin-section work relating to when particular developmental stages occur during the lifetime of dinosaurs has been done on some of the ones from Transylvania. Ragna Redelstorff, Zoltán Csiki, and Dan Grigorescu conducted histological studies of a series of femora from *Telmatosaurus transsylvanicus*, to test whether the largest among them (464 mm long) come from small adults (i.e., dwarfed taxa) or from juveniles of more normal-sized taxa.[52] Histologically, these femora exhibit up to eight lines of arrested growth, and the outermost layer of the largest femur is very thin and avascular, indicating that growth had either slowed down considerably or ceased. Such bone microstructure indicates that the largest femora belong to fully adult dwarfed individuals, not juveniles of much larger forms. The reduction in body size was presumably caused by a slowdown of their growth rate.

A decidedly different bone histology is encountered in *Magyarosaurus dacus*.[53] Thin-sections were again made of different sizes of long bones (femora 346–545 mm long, humeri 223–488 mm long) but, unlike the situation in *Telmatosaurus*, the overwhelming majority of the cortex in each of these sections is extensively remodeled by secondary bone. There is almost no preservation of primary bone that could record lines of arrested growth, even in the smallest individuals. Koen Stein and his coworkers noted that the intense secondary remodeling in the bones of *Magyarosaurus* is closest to that of other individual sauropods of very late histological ontogenetic stages. Although lacking a thin avascular outer layer that would indicate reduction or cessation of growth (as in *Telmatosaurus*, above), the early appearance of secondary remodeling in *Magyarosaurus* suggests an early onset of sexual maturity and a short lifespan.

Finally, there is *Zalmoxes robustus*. As before, with *Telmatosaurus*, Redelstorff and her coworkers studied the histology of a series of femora (164–355 mm long) to test whether the largest of these represent dwarfed adults or juveniles of normal-sized taxa.[54] The long-bone histology reveals a slow growth rate in *Zalmoxes*, indicated by the high number (15) of narrow-spaced lines of arrested growth. However, vascular canals in the cortex open onto the bone surface, which indicates that growth had not plateaued at the time of death. Thus it is likely that the existing material of *Zalmoxes* represents bones not yet fully grown, and this ornithopod can probably be characterized by an extreme slowdown of its growth rate and an extended growth period.

We began this chapter describing the ways in which dinosaurs have pushed the envelope of body size, and, in so doing, they have provided a heterochronic context for their enormity. Although peramorphosis appears to be the main driving force in the evolution of nonavian dinosaurs,[55] several examples of paedomorphosis have also been identified in this group,[56] and these claims have led us to consider the importance of heterochrony more generally in dinosaurian evolution. Nopcsa's notion that the dinosaurs from Transylvania were dwarfed could well be equivalent to seeing them as an assemblage of heterochronic Peter Pans. To this end, we have begun exploring both ontogenetic and phylogenetic aspects of both size and shape changes in the hadrosaurid *Telmatosaurus*, as well as in the titanosaurian sauropod *Magyarosaurus* and the primitive ornithopod *Zalmoxes*.

Who, What, Where, and When: Optimization and Cladograms

Optimization is our tool of choice when placing characters in their phylogenetic context. Once we have an explicit, well-supported cladogram for a particular group of organisms, we can then use this tree to determine the most parsimonious distribution of other aspects of this taxon, be it adaptation, coevolution, historical biogeography, biomechanics, or ecology. The use of phylogenetic trees as the basis for determining the most parsimonious sequences of evolutionary transformations for these additional characters—known as character optimization—does not help build trees, but instead is used for evaluating the behavior of other features on trees that are already constructed.[57] This *a posteriori* method has been used to evaluate historical biogeographic relationships, to infer soft-tissue anatomy in extinct vertebrates, and to project the function and behavior of living organisms into the past.[58]

Let's look at some simple examples of how *a posteriori* character optimization works. Figure 5.6a illustrates four taxa with a known relationship, all of which also share the same character (here we go with geographic distributions—they are all known from area A—but the characters to be optimized might be soft anatomy, biomechanics, physiology, or other features not included in the original cladistic analysis). Where is the ancestor of taxon 1 and taxon 2 (node 5) likely to have come from? Of course, it could be anywhere, but the most parsimonious interpretation is that this ancestor shared the same locale (area A) as its two descendants. When further relationships are considered (taxa 3, 4), since they all come from area A, then it's obvious that the ancestors (nodes 6, 7) are also inferred to have come from area A. Once we're at the base of the tree (character generalization), we then, in figure 5.6b, double check our ancestral characters back up the tree to resolve any ambiguities that may be left (character optimization). In this case, there is none—all of the nodes have A as their ancestral area. In other words, the evolution of this clade would be considered to be endemic to area A.

Yet what if there is a mixture in where the members of a clade come

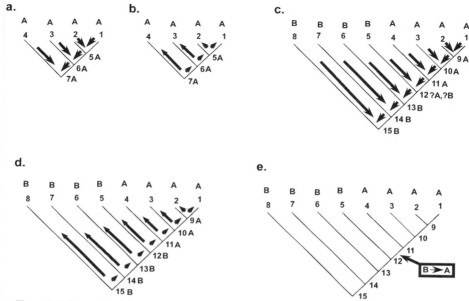

Figure 5.6. A posteriori character optimization (see text for explanation)

from? In figure 5.6c, half of the taxa (1–4), those to the right, are from area A, while those from the other half (5–8), on the left, are from area B. Generalizing down the tree, we know that the ancestors reconstructed at nodes 9, 10, and 11 would also have been found in area A, using the same logic we applied when confronting the situation in figure 5.6a. But what about the ancestral area at node 12—is it area A or B? Clearly something happened here, but we don't yet know what it is. In truth, it could be either one, so we register it as "?A / ?B" and then move down the tree yet another step to node 13. Here, we compare the area for taxon 6 and that for node 12. Area B is held in common with node 12, so we infer that the ancestral area at node 13 is B. Thereafter, down the tree, area B is inferred to the ancestral areas at nodes 14 and 15.

In order to resolve what's going on at node 12, in figure 5.6d we optimize the characters back up the tree. The ancestral area at nodes 15, 14, and 13 is clearly B. We then compare node 13 with taxon 5 to resolve the ambiguity that we encountered during the generalization phase at node 12. Area B is held in common between node 13 and taxon 5, so we infer that the ancestral area at node 12 is B. With all remaining ancestral areas already identified as A, figure 5.6e shows that the shift from area B to area A occurred between nodes 12 and 11.

These examples have been simple and direct, but you can imagine how strenuous it would be to do this by-hand effort for clades with many members from disparate geographic regions. This problem is even more acute for sets of cladograms that are equally the most parsimonious ones. Fortunately, *a posteriori* optimization of characters on a tree is more easily accomplished by using any of the available numerical cladistic algorithms, such as PAUP, MacClade, and Winclada.[59]

Getting Small, *Telmatosaurus*-Style

Combing through the drawers of fossil bones and teeth, often in dusty, dark basement rooms in museums and universities, is very much like being in the field. You never know what you might discover next. Perhaps it will be some bone that indicates something new, something different, or something overlooked by the people who had studied it before.

Imagine yourself inside the Hungarian Geological Survey building, with its picturesque blue-and-white checked ceramic roof and rooms with vaulted ceilings, being turned loose among the collection of fossils by Dr. László Kordos, curator of these bones and teeth. Here, amid the pleasant smells of pipe smoke and hearty coffee, is where the considerable fossil remains from the Late Cretaceous of Transylvania are stored.

Our trek through these specimens yields the usual assortment: isolated vertebrae and limb bones, trays of teeth, and indeterminate scraps of bone. No surprises there—we've seen the likes of these before, and it's easy to assign them to particular kinds of dinosaurs. Then, from one of the drawers, we find something totally unexpected. According to the label, these teeth are supposed to belong to *Telmatosaurus*, but they certainly don't look like any sort of hadrosaurid teeth we've seen before. They're relatively wide, somewhat like the teeth of *Iguanodon*, and they don't display the typical strongly developed, vertical ridge of other hadrosaurids; instead, there are several low ridges on the enameled side of each tooth (figure 5.7, middle). Excitedly, we think we've discovered a new form of dinosaur, previously overlooked by Nopcsa and all who have come after him. At least that's our first reaction. However, we must be sure whereof we speak, so it's off to the literature. There, in black-and-white on plate VI, figure 3, in Nopcsa's 1900 monograph on *Telmato-*

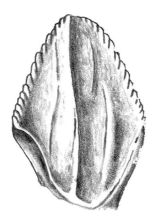  

Figure 5.7. The lower teeth of *Iguanodon* (*left*), *Telmatosaurus* (*middle*), and *Maiasaura* (*right*). Scale = 1 cm

saurus, are these same teeth: he had clearly recognized that they belonged to this hadrosaurid and, furthermore, that this lower dentition had a different morphology than do the teeth in the upper jaws.[60] Disappointed and chastened, we are nevertheless enlightened and perplexed at the same time. What are these teeth of *Telmatosaurus* telling us?

In order to answer this question, we need to look at the teeth of a wide array of iguanodontian ornithopods, such as *Iguanodon*, *Ouranosaurus*, *Camptosaurus*, and *Tenontosaurus*, and try to understand how they changed during the process of tooth replacement. Morphologically, the squat, lozenge shape of the teeth of these forms is regarded as the primitive condition for the Iguanodontia. As these ornithopods grow during their lifetime, their "baby" teeth are rather small, but, as these teeth are replaced throughout life, they become proportionately much larger. Concomitantly, the number of teeth necessary to fill up the jaws doesn't increase very much, compensating instead by an increase in tooth size during ontogeny. Take *Iguanodon*, for example (figure 5.7, left). As it grows, its teeth get much larger as they are replaced (a 300% increase in tooth size during ontogeny), but the number of tooth positions remains roughly the same (about a dozen tooth positions from juvenile to adult).

Compared with this more primitive system of replacement, hadrosaurids do something entirely different. In these ornithopods, the dentary teeth increase in size only slightly during ontogeny. With adult teeth

Figure 5.8. Telmatosaurus as a dwarf. The solid silhouette represents an iguanodontian ornithopod of ancestral body size; the open silhouette represents *Telmatosaurus*. Scale = 2 m

nearly the size of the "baby" teeth, it is the enormous increase in the number of tooth positions in the jaws that compensates for increasing jaw size as the animal gets larger during ontogeny. For example, *Maiasaura* hatchlings have a dentary dentition consisting of approximately 10 tooth positions filled by 6 mm wide teeth, whereas adults have jaws containing 40–45 tooth positions filled by 8 mm wide teeth (figure 5.7, right). Compared with the primitive ornithopod condition, hadrosaurids have added teeth like crazy, expanding not only the length of the dentition as the jaw grows, but also providing the possibility of having a greater number of closely packed replacement teeth at one time. In this way, the classic hadrosaurid dental battery is created.[61]

Fitting *Telmatosaurus* into this picture, our Haţeg hadrosaurid is little different morphologically from the somewhat more basal iguanodontians *Ouranosaurus* or *Iguanodon*, except in two respects—its smaller body size (figure 5.8) and the size and shape of its teeth (figure 5.9). The upper teeth of *Telmatosaurus* are narrow, diamond-shaped, and equipped with a single, centrally placed ridge; in other words, they are most like the juvenile condition seen in non-hadrosaurid iguanodontians. Its lower teeth, in contrast, are wider, asymmetrical, and bear several low ridges, making them intermediate between those of other hadrosaurids and more primitive iguanodontians. These teeth, too, were small, but they most resemble the shape of adults of non-hadrosaurid iguanodontians.

How this baby-toothed condition arose in hadrosaurids can be evaluated by optimizing body size and tooth development onto the cladogram of Iguanodontia (figure 5.10). Optimization is an *a posteriori* method for

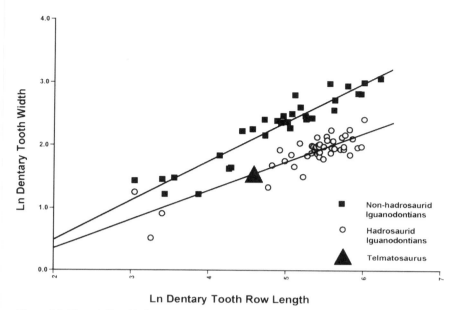

Figure 5.9. The relationship between dentary tooth width and tooth row length for both non-hadsaurid iguanodontians and hadrosaurids. Note that *Telmatosaurus transylvanicus* plots among juvenile hadrosaurids and not far from the juveniles of non-hadrosaurid iguanodontians.

identifying the most parsimonious sequence of character changes by reference to an explicit phylogenetic tree. As we learned in chapter 2, the majority of hadrosaurids belong to two major subgroups, the hollow-crested lambeosaurines and the flat-headed, nasal-arched, solid-crested hadrosaurines, which together form the group called Euhadrosauria. Several hadrosaurids, *Telmatosaurus* prominent among them, are positioned below this clade of euhadrosaurians, but above non-hadrosaurid iguanodontians such as *Ouranosaurus* and *Iguanodon*.[62] When body and tooth size are optimized on this cladogram, they fall out as a peramorphocline from the base of Iguanodontia to the base of Hadrosauridae. At this nexus, something happens to this ever-increasing relationship: body size trends reverse themselves, and the basal members of the hadrosaurid clade then undergo downsizing, not only in body size, but also in the size of their teeth. Thereafter, the evolutionary history of tooth size is decoupled from that of body size—as body size trends become peramorphic again, the teeth remain miniaturized, like those of juveniles.

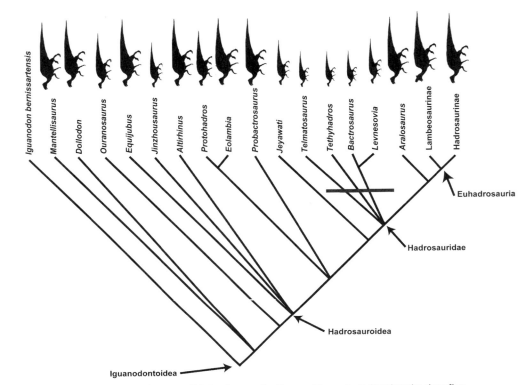

Figure 5.10. A cladogram of higher iguanodontian ornithopods, indicating the dwarfing event leading to *Telmatosaurus* and *Tethyshadros*. The bar immediately below Hadrosauridae indicates the acquisition of miniaturized maxillary teeth.

In summary, we regard the small adult size seen in *Telmatosaurus* as the trigger for the evolution of the dental battery of duck-billed dinosaurs, the hallmark of Hadrosauridae. By arriving at adulthood having a smaller body size than its ancestors did, *Telmatosaurus* was assured that its teeth would resemble those in the younger ancestral stages. This juvenilization of their teeth amounts to arrested growth, otherwise known as paedomorphosis. Thereafter, euhadrosaurians increased in body size, but retained their juvenile dentitions into adulthood. This juvenile dentition, which was organized into a closely packed battery of miniaturized teeth with a complex occlusal surface, was passed on to their descendants.[63] In this way, the Peter Pan syndrome provided the means by which the duckbills developed their complex dentitions.

Magyarosaurus: The Smallest of the Largest

Being the largest of all terrestrial organisms, sauropods assuredly packed a lot of weight onto their limbs. Standing still, an average adult *Apatosaurus* had to withstand 10 metric tonnes on its forelimbs and 24 metric tonnes on its hind limbs and, to amble across the Mesozoic countryside, these same limbs would have to have been built to withstand double or triple these loads.[64]

The stresses on the limbs of sauropods smaller than *Apatosaurus* obviously would have been less. This would also have been the case for juveniles and subadults of *Apatosaurus* and of youngsters of other species. Does that mean that the limb bones of adults should display different scaling properties when compared with those of a growth series? Because *Magyarosaurus* adults are the smallest among sauropods, we were interested to find out.

At a length of 5–6 m, and weighing in excess of 750 kg, *Magyarosaurus* is certainly among the smallest of all known sauropods (figure 5.11), as originally noted by Nopcsa in 1915.[65] Although there are no complete or even partially articulated skeletons, this sauropod is known from abundant material representing nearly the entire skeleton, all collected from the Upper Cretaceous rocks of Transylvania.

In addition to gaining a reasonably good understanding of the anatomy of *Magyarosaurus*, we are also coming to learn more about its position in the evolutionary history of sauropods. Since their original discovery and the fulsome years of exploration in the American West, the excavations at Tendaguru (Tanzania), the finds in Sichuan (China), and the discoveries from Patagonia in South America, the phylogeny of sauropods was anything but clear. However, thanks to a number of recent phylogenetic analyses, their internal relationships are becoming much better understood.[66] For our purposes, it is important to remember that *Magyarosaurus* is a titanosaur, a clade of sauropods that includes *Rapetosaurus*, *Saltasaurus*, and *Malawisaurus*, and it is more closely related to *Brachiosaurus* and its relatives than it is to other sauropods such as *Diplodocus* (chapter 2). Together, all of these sauropods play an important role in providing the evolutionary context for understanding the heterochronic significance of *Magyarosaurus*.

In the late 1990s, the two of us, along with Jason Mussell, then a graduate student at the Johns Hopkins University, decided to approach

Figure 5.11. Magyarosaurus as a dwarf. The solid silhouette represents a titanosaurian sauropod of ancestral body size; the open silhouette represents *Magyarosaurus*. Scale = 3 m

growth and development in these sauropods by examining the size and shape of the humerus and femur. We collected a variety of measurements from our specimens and analyzed the significance of these data, using both statistics and phylogeny. For our statistical analyses, we assembled one series consisting solely of adults and another that combined growth stages drawn from a mixture of 25 titanosaurs and brachiosaurs that are closely related to *Magyarosaurus*.[67] By focusing on the long bones of these sauropods, we sought patterns of heterochrony using young and adult individuals, in this case in terms of the biomechanics of long-bone design. A limb bone must obviously be built to withstand, without fracturing, the great loads that pass down their length. A bone's ability to resist this kind of load is proportional to the area of its cross-section, so for sauropods, long-bone proportions are particularly sensitive to changes in body size. Naturally, we wanted to see just what kind of humeral and femoral dimensions *Magyarosaurus* and other sauropods, young and old, have relative to each other, by examining them in a way that reflects their strength. This again is a exercise in scaling, so we set about determining the relationship between the length of these long bones and their midshaft widths, dividing the sauropod sample into individuals thought to be fully adult, and other individuals that could be assembled into a growth series from smallest to largest. Age classing of humeri and other bones is often difficult with ever-growing animals such

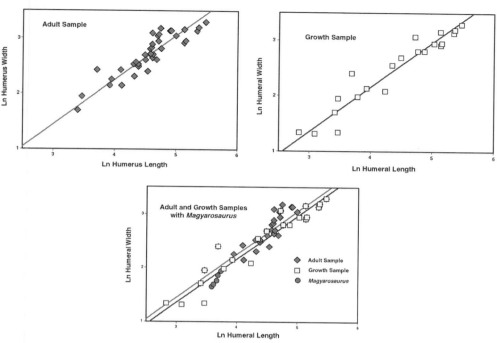

Figure 5.12. The relationship of humeral length and width (midshaft diameter) in titanosauroid sauropods. The interspecific regression of adults is indicated in the graph on the upper left. The regression of the composite growth series is indicated in the graph on the upper right. The combined adult and growth regressions—with *Magyarosaurus* superimposed on these lines—are indicated in the graph at the bottom.

as dinosaurs. However, as we have already seen in the case of *Telmatosaurus* and the other ornithopods we compared it with, it was possible to get an idea about the approximate maturity of an individual sauropod. The features we looked at were the degree of fusion of the bones of the braincase and vertebrae, the maturity of the joint surfaces of the limb bones, overall body size, and (sometimes) anecdotal comments in the literature. In nearly all cases, the largest individuals of a species were identified as adults.

What we report here comes from our analyses of the humerus, although roughly the same applies to the femur. When we plotted the data from the sample composed exclusively of adults, we got a reasonable linear relationship between humeral length and midshaft diameter (figure 5.12). Then, from our assembled growth series, we obtained another

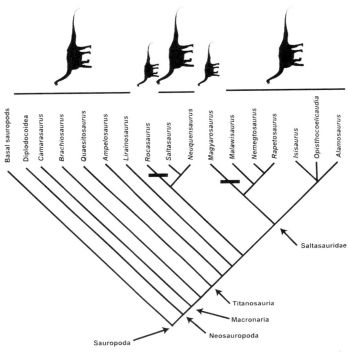

Figure 5.13. A simplified cladogram of titanosauriform sauropods, showing two dwarfing events—one in *Magyarosaurus* and another in *Rocasaurus*—as indicated by the bars. (Modified after Curry Rogers and Forster 2001)

line with a similar slope to that of the adult sauropods, but which plotted slightly beneath the latter.[68] This means that the humerus of a growing sauropod would be narrower than the same length of humerus for an adult. We then added data from *Magyarosaurus* specimens, and put a few other titanosaurs to the mix, to see how closely they resembled either the adult or the growth samples. *Magyarosaurus* plotted closest to the growth-series line, rather than to the adult line. Beyond mere inspection, the *Magyarosaurus* data are also statistically more similar to the growth series than to the adult sample.[69] Therefore, we concluded that *Magyarosaurus* humeri appear to be more similar to those of subadults than to adults of other taxa.

To judge whether these statistical insights had a phylogenetic context, we optimized body size at adulthood onto a cladogram of *Magyarosaurus* and its fellow titanosaurs.[70] Figure 5.13 represents one such cladogram of these sauropods.⁷¹ Here we can see that small size evolved in the lineage leading to *Magyarosaurus* (as well as that leading to *Rocasaurus*). As did *Telmatosaurus*, *Magyarosaurus* joins Peter Pan's merry band by retaining juvenile features into adulthood—that is, by paedomorphic dwarfing.

Was *Zalmoxes* a Dwarf?

At approximately 3–5 m long, *Zalmoxes* does not strike one as an extremely small iguanodontian dinosaur (figure 5.14). Nonetheless, the more relevant questions are: Is *Zalmoxes* bigger or smaller than its closest known relative, *Rhabdodon* from the Late Cretaceous of France and Spain? How do these two ornithopods compare with their next closest relatives, the iguanodontian ornithopods? And thereafter?

Zalmoxes certainly doesn't appear to be especially different in its proportions from most other basal iguanodontians. The skull isn't particularly shorter than that of *Tenontosaurus*, and the details of its construction don't cry out for heterochronic interpretations. However, the postcranial skeleton appears to be more robust and heavy, and we would have been remiss if we didn't pursue all heterochronic connections related to changing body size and skeletal proportions. Consequently, we needed to look at how the likes of *Zalmoxes*, *Rhabdodon*, *Tenontosaurus*, *Orodromeus*, and *Hypsilophodon* grew.

We approached growth and development in these ornithopods much as we had with *Magyarosaurus*, although we chose to examine only the femur, since it was much better represented in our sample of bones than the humerus.⁷² Measuring the length and midshaft width of these femora, we then analyzed the significance of this set of data, again using both statistics and phylogeny. However, unlike our analyses of *Magyarosaurus*, where we had to cobble together a single, generalized growth series for titanosaurs, virtually all of the ornithopods that we wanted to compare with *Zalmoxes* are known from adults, subadults, and even juveniles, so we had direct information on the growth patterns of each of these taxa. The other significant difference from our *Magyarosaurus* analyses was that we had a much better understanding of the phylogenetic history of ornithopods than we did for titanosaurs and other sauropods (chapter 2). This, in fact, provided us with a detailed, well-tested template for the comparisons we wanted to make.

Figure 5.14. Zalmoxes *as a dwarf. The solid silhouette represents a basal iguanodontian ornithopod of ancestral body size; the open silhouette represents* Zalmoxes robustus. *Scale = 3 m*

We began our work by rounding up as large a sample as possible of skeletal elements for both species of *Zalmoxes*. We selected the femur (as noted above), which, in the case of *Z. robustus*, turned out to represent samples from seven individuals with femora distributed from about 20 to 40 cm in length, and, for *Z. shqiperorum*, three individuals with femora ranging from 28 to 45 cm. Dimensions measured from several femoral landmarks were then used to produce growth curves for the *Zalmoxes* sample. The same data were collected for a selection of ornithopods that are thought to be closely related to *Zalmoxes*: *Rhabdodon*, *Tenontosaurus*, *Orodromeus*, and *Hypsilophodon*. These growth curves were then compared pairwise with each other.[73]

The overall pattern of growth for *Zalmoxes*—treated here in terms of both species (*Z. robustus* and *Z. shqiperorum*)—and for closely related ornithopods is quite similar (figure 5.15). Indeed, when we compared growth curves between any two of these ornithopods (say, between *Orodromeus* and *Tenontosaurus*), they were statistically indistinguishable from each other. The same applies to comparisons involving the two species of *Zalmoxes*, as well as comparisons with *Rhabdodon*—their growth curves are virtually identical. This overarching trend among these ornithopods indicates that the femur changed its proportions in the same fashion, regardless of whether it belonged to *Hypsilophodon*, *Tenontosaurus*, or *Zalmoxes*. The largest individuals, then, constitute the only difference among the samples. For the biggest *Tenontosaurus* or *Rhabdodon*, the femur grew to a length of nearly 60 cm. Right behind that comes *Zalmoxes shqiperorum*, with a femur length of just over

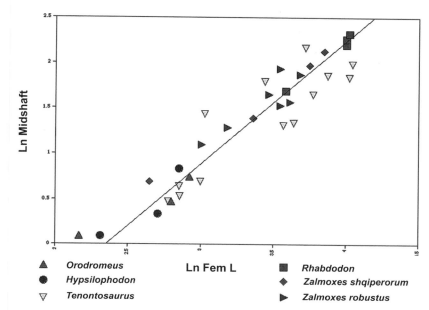

Figure 5.15. The relationship between femoral length and midshaft diameter in basal ornithopod dinosaurs. Included are growth series obtained from *Orodromeus makelai*, *Hypsilophodon foxii*, *Tenontosaurus tilletti*, *Rhabdodon priscus*, *Zalmoxes shqiperorum*, and *Z. robustus*. The line represents the regression of femoral length and midshaft diameter for the combined sample of all species.

45 cm. *Z. robustus*, with a femur length of 37 cm, is smaller still, whereas *Orodromeus* and *Hypsilophodon* have the shortest femora (both between 15 and 20 cm long).

As far as this simple scaling analysis goes, both species of *Zalmoxes* appear to grow their femora just like other ornithopods. One of them (*Z. shqiperorum*) does so to the degree exhibited by the existing sample of *Tenontosaurus*, but the other (*Z. robustus*) is known from femora that are greater in length than those of *Orodromeus* and *Hypsilophodon*, but smaller than those of *Rhabdodon*, *Tenontosaurus*, and *Z. shqiperorum*. Is this pattern one of successive peramorphosis in the clade leading to *Tenontosaurus* and the larger species of *Zalmoxes*, or of paedomorphosis leading to *Orodromeus* and *Hypsilophodon*? Or is it a mixture of both? Without full-grown adults from this taxon (see the discussion above on bone histology), we really can't tell when *Zalmoxes* stopped

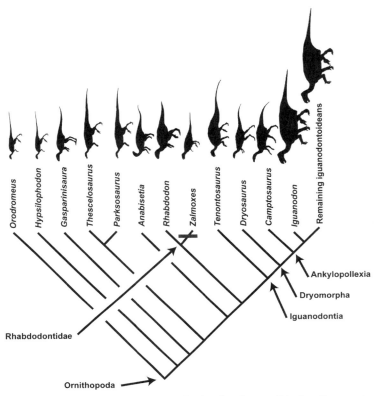

Figure 5.16. A cladogram of basal ornithopods, showing the possible dwarfing event leading to *Zalmoxes robustus*, as indicated by the bar

growing. This ornithopod slowed down its growth rate and had an extended growth period, but it had not yet achieved its largest size. To see what this eventuality entailed in a phylogenetic context, we again optimized these size data onto the cladogram for basal euornithopods (figure 5.16). We can see that the more primitive trend within this part of the clade consists of a peramorphocline from basal euornithopods such as *Orodromeus* through more highly positioned taxa like *Hypsilophodon*, *Tenontosaurus*, *Rhabdodon*, and *Zalmoxes shqiperorum*. However, downsizing appears to occur in *Zalmoxes robustus* within the clade represented by the two last-mentioned taxa, all other thing being equal. If this is true, then we may have another Peter Pan paedomorph in *Z. robustus*.

Struthiosaurus, Theropods, and the Rest

Positive evidence exists for dwarfism in *Telmatosaurus, Magyarosaurus*, and perhaps *Zalmoxes robustus*. What of the other members of the Transylvanian fauna? The ankylosaur *Struthiosaurus* and the unnamed pterosaur are quite a bit smaller than expected, compared with their phylogenetic compatriots, but we're not yet certain whether they are represented by mature adult material or by subadult specimens. In contrast, the dromaeosaurid and troodontid theropods, although small, are no smaller than their close relatives in Asia and North America. Only *Hatzegopteryx*, poorly known though it may be, is large, but with an estimated wingspan of 12 m, it may turn out to be among the largest of all the azhdarchid clade.[74] Turning to the Transylvanian crocodilians, turtles, and mammals, they are neither more diminutive nor more sizeable than what would be expected, based on their evolutionary cohorts. Thus this triumvirate of turtles, mammals, and crocodilians don't seem to be playing the heterochrony game.

In the end, then, we have within the Transylvanian fauna a handful of cases of dwarfing through paedomorphosis, a roster of normal-sized animals, and an example of gigantism by way of peramorphosis. Maybe you've been expecting universal dwarfing among the animals found in Transylvania, but why should all of these forms have gotten small? We're more than happy that paedomorphic downsizing can be identified in as many Transylvanian groups as it has—for here we have an important evolutionary signal. Nopcsa knew that such body sizes, compared with those elsewhere in the world, flew in the face of the basic tenets of dinosaur evolution, where large size reigns supreme, so he sought a causal basis for all of the changes in body size among dinosaurs.

THE BIOLOGY OF GETTING BIG AND SMALL

As Nopcsa saw it, the evolution of both large and small dinosaurs came about through some sort of neo-Lamarckian mechanism. For him, this inheritance of acquired characteristics was mediated by diseases or other physiological disturbances.[75] For example, he looked to such endocrine diseases as acromegaly, in which hyperfunction of the pituitary gland leads to gigantism and a thickening of the bones in humans, to account for an increase in body size. His evidence for acromegaly was the enlarged pituitary fossa found in dinosaur braincases.

Other explanations of dinosaurian gigantism, even up to the 1950s, took a decidedly non-Darwinian perspective. Otto H. Schindewolf, one of the twentieth century's great macroevolutionary paleontologists, devoted much of his long career to defending his theory of typostrophy. Schindewolf sought to understand evolutionary patterns as being internally regulated: groups of organisms, like individuals, go through phases similar to birth, childhood, adulthood, old age, and death as they evolve, and their extinction therefore is inevitable, predicated on this "evolutionary lifecycle."[76] In dealing with dinosaur extinction, he argued that these animals were driven to a sort of "racial senescence" that accompanied their growth to an exceedingly large size, as were other organisms, such as ammonites. How else to account for the "inevitable" poor design of dinosaurs? "It's quite likely that these gigantic forms were no longer able to carry their enormous body weight around on land. They lived like amphibians, submerged in swamps and lagoons where water lightened the load and made locomotion easier."[77]

The then-popular notion of dinosaurs as evolutionary failures because of their extinction at the end of the Cretaceous masked, for Schindewolf, just how these animals attained their gigantic size. He was forced to search for an ecological context, in which these animals were in their senescent phase. We now know that dinosaurs were not particularly relegated to the swamps, nor do taxa have well-defined "life stages," but Schindewolf was right to seek evolutionary and ecological explanations for the changes in body size in dinosaurs. For example, their large size has been related to their physiology, in particular to the ways in which these animals regulate temperature: either through mass homeothermy (stable internal temperatures due to their large body's low surface area / volume ratio) and endothermy (warm-bloodedness based on metabolic heat generation).[78]

Size increases through phylogeny not only accompany dinosaur evolution, but are also notoriously common elsewhere in the fossil record, so much so that they have been generalized under the rubric of Cope's Rule, named after the prominent American neo-Lamarckian evolutionary biologist and vertebrate paleontologist Edward Drinker Cope (1833–1897).[79] Ironically, its counterpart, evolutionary dwarfism, has received much less attention and is mostly centered on insular dwarfing, where changes in body size—downsizing in particular—are very common.[80]

Nopcsa viewed dwarfing as a consequence of body-size effects through island habitation; in other words, Transylvanian dinosaurs evolved to a smaller size as they became geographically isolated on an island in the Neotethyan Sea. There is ample indication that Transylvania, at the end of the Cretaceous, was a remote terrestrial region, a relatively small, isolated landmass revealed by the receding sea. This insular environment surely must have been large enough to accommodate viable populations of a variety of dinosaurs and other vertebrates, but apparently it was not so large as to support sauropods and ornithopods, at least in their normal, quite large body sizes.

We hope by now that it is obvious that the Transylvanian fauna includes a number of dinosaurs of small adult body size. The region in which these diminutive forms lived was an insular habitat, which leads us (in the same way as Nopcsa) to look for an explanation for our dwarfed Transylvanian dinosaurs by examining the relationship between body size and isolation.

The pattern of island dwarfing among mammals has been termed the Island Rule by Chicago evolutionary biologist Leigh Van Valen, who noted that it "seems to have fewer exceptions than any other ecotypic rule in animals."[81] The epitome of the Island Rule must assuredly be the dwarfing of insular elephants commonly present in the Pleistocene. These diminutive proboscideans are interesting—and relevant—for two reasons. First, among animals with living relatives, they most closely approximate dinosaurs in terms of body mass. However, even more importantly, their pattern of dwarfing is well enough understood, thanks to V. Louise Roth, for us to explore our own patterns of dwarfing in Transylvania. Roth, an evolutionary biologist at Duke University, conducts research on both large and small elephants, particularly in terms of their anatomy, feeding habits, growth, ecology, reproduction, fossil record, and evolution. Armed with this information, she developed a multifaceted proboscidean model that relates interspecific competition, predation, and subsequent population densities, in order to explore what is causally behind the Island Rule.[82]

Roth's model begins by taking up the events of island colonization. Getting to these islands is clearly a function of their remoteness from the adjacent mainland, and surviving such a migration depends on a combination of good luck and preparedness. In the case of nonhuman migrants, preparedness is a matter of body size. As it turns out, large individuals are more likely to be successful in surviving overseas dispersal and colonization than small individuals; hence, the founding population on an island may be of relatively large body size.[83] This larger initial body size may be counterintuitive, but it provides a means for surmounting the problems of breaking through the barrier of isolation.

At this point, the largish elephants of Roth's model have survived dispersal and are now island residents. Consider, though, that this new home base for the elephants must be much more confining than their previous range on the mainland. With this more restricted land area comes an ever-greater exploitation—and the eventual deterioration—of existing food supplies as the isolated populations reach high densities and the elephants suffer from increased overcrowding. Smaller body size is favored under these conditions and, according to Roth's model, is most likely to arise when individuals reach reproductive maturity at earlier—and therefore smaller—stages. This connection between reproductive maturity and body size among dwarfed insular elephants again focuses our attention on the role of heterochrony and, in particular, on paedomorphosis. This shift of sexual maturity to younger and therefore smaller individuals is the most important connection with individual fitness.

So far, so good—the dwarfed elephant model provides some similarities with the Transylvanian dinosaurs. The paleogeography of the Tethyan realm during the Cretaceous indicates that terrestrial habitats were relatively well isolated from each other and often limited in area. Thus conditions in Transylvania are consistent with the geographic and ecological underpinnings of the elephant model.

We likewise argue that small size in *Telmatosaurus*, *Magyarosaurus*, and perhaps other Transylvanian forms appears to be favored under conditions of decreasing food supply, due to increased population densities in restricted habitats, just as it appears to be the case in the dwarfed, island-dwelling elephants. In this way, the dwarfing of Transylvanian dinosaurs would have arisen through paedomorphosis: individuals reaching reproductive maturity at earlier, smaller stages.

From a faunistic perspective, the dinosaurian herbivores are dwarfed relative to their close relatives elsewhere in the world, whereas the few predators known from Transylvania are not much different in body size from their closest relatives. In terms of the elephant model, these observations are not surprising. The lack of large individuals among the herbivores strongly suggests that immigration to this restricted area must have taken place considerably earlier than the latest Cretaceous. (The time interval for isolation and colonization of the Transylvanian region will be discussed further in chapters 6 and 7.)

As for body size among the Transylvanian theropods, as we have indicated, there were no behemoth, *Tyrannosaurus*-like predators living in this region in the Late Cretaceous. This, too, is not a surprise. The isolated habitats of the kind we envision for Transylvania were probably capable of supporting a lesser quantity and diversity of predators, mainly ones that had smaller home ranges; and, of these, those with smaller bodies fit the more restrictive Transylvanian ecological conditions better than large-bodied predators.

The downward trend in body size among large herbivores, which Roth's model links with limited island resources, may also be driven by differences in predation between the mainland and an island. We know that large mammalian predators, though present on the mainland, were absent from islands inhabited by dwarfed elephants during the Pleistocene, and the same appears to have been true for most of the dinosaurian predators of the ancient Transylvanian region.[84] Lower levels of predation on islands may act as a permissive factor, allowing large herbivores to evolve into smaller, perhaps more optimal sizes, since large size would not be selected as a deterrent to predators. In addition, this rationale for insular dwarfism may also help explain the evolution of gigantism among many small mammals and other animals on islands. Take, for example, the case of the modern giant rats on the Lesser Sunda Islands in the South Pacific.[85] Such rat gigantism—a nearly half meter long rodent, without the tail that gives new meaning to *The Princess Bride*'s "rodent of unusual size"[86]—may be viewed as evolving closer to its optimal body size in the absence of the usual roster of predators from the mainland. Similar to the counter-case of island downsizing among large animals, very small size is no longer necessary as a defense against predators, producing, as a consequence, an increase from a smaller to a larger body size.[87] This embellishment to the Island Rule provides an argument for there being an optimal intermediate body size that organisms would be free to evolve toward when predator pressures are significantly altered, as they would be on an island.[88]

Little Giants and Big Dwarfs

Whether dwarfism in *Magyarosaurus*, *Telmatosaurus*, and maybe *Zalmoxes* was spurred by limited resources or by an alteration and perhaps a reduction in predation, these small dinosaurs beg for a heterochronic and ecological explanation. If Roth's elephant model applies to the Transylvanian dwarfs, and we believe that it does, the adaptive significance of dwarfing is its immediate relationship with life-history strategies such as accelerated sexual maturation (i.e., paedomorphic heterochrony), rather than it being a development based on morphological advantages alone. As Gould put it, the "timing of reproduction is an adaptation in itself, not merely the consequence of evolving structure and function."[89] The morphological consequences of paedomorphosis—juvenile morphology—come along initially as biological baggage: a consequence of downsizing due to isolation. Thus, for example, it is not the retention of a juvenile dentition and the formation of a dental battery in *Telmatosaurus* that is the primary focus of selection. Instead, the overarching selection is for a small body—that is, paedomorphic dwarfism, based on the ecological particulars of the isolated region of Transylvania.

CHAPTER 6

Living Fossils and Their Ghosts

*Being a Short Interlude on Coelacanths
and Transylvanian Ornithopods*

For evolutionary biologists of all sorts, 23 December 1938 was a very important day. It was then that a trawler called the *Nerine* put into port at the town of East London, located about 850 km to the east of Cape Town, South Africa. The skipper, Captain Hendrick Goosen, made a living fishing the nearby coastal waters of the Indian Ocean. Having made friends with Marjorie Courtenay-Latimer, the curator of a small local museum, he would often have the dockman call Courtenay-Latimer to come look over the *Nerine*'s catch and to take any unusual specimens she wanted for her museum. On this particular day, the *Nerine* entered port after trawling off the mouth of the nearby Chalumna River. When the dockman called Courtenay-Latimer, she took a taxi to the ship, delivered her Christmas greetings to the captain and crew, and went through that day's catch for anything unusual. There, beneath a pile of rays and sharks on the deck, was, as she put it, "the most beautiful fish I had ever seen, five feet long, and a pale mauve blue with iridescent silver markings."[1] Although she had no idea what the fish was, she did know that it had to go back to the East London Museum at once. After convincing the taxi driver to allow the reeking, 5 ft fish to accompany her, he drove her and the specimen back to the museum.

Back in her office, Courtenay-Latimer tried to identify the bizarre creature (figure 6.1) by combing through the few reference books in her library. A picture of a long-extinct fish bore the greatest resemblance, particularly in the structure of the head and the trilobed shape of the tail. She made a crude sketch of her discovery and sent the drawing and a short description to Professor J. L. B. Smith, a chemistry professor at Rhodes University in nearby Grahamstown, who also was locally well known as an amateur ichthyologist. Unfortunately, Smith was away for the Christmas holidays and the consensus back home was going against her increasingly odiferous fish—the director of the East London Museum dismissed it as a common rock cod. In an effort to preserve the fish by mounting it, much of the viscera had been discarded and then lost, and, to the great disappointment of all concerned, even the photographs taken of the preparation were spoiled.

When Smith finally visited the East London museum on 16 February, he immediately identified the fish as a coelacanth, a group of fish thought to have gone extinct toward the end of the Cretaceous, some 80 million years ago. Called the "most important zoological find of the century," this discovery made Smith and Courtenay-Latimer overnight celebrities. In 1939, the fish was named *Latimeria chalumnae* (after Courtenay-Latimer and the river near where it had been collected) by Smith.

The saga of the coelacanth continues. On 21 December 1952, Captain Eric Hunt, a British sailor operating in the waters off the Comorean island of Anjouan, was approached by two islanders carrying a hefty bundle. One of them, Ahamadi Abdallah, had caught a heavy, grouperlike fish by hand line, while the other—a schoolteacher named Affane Mohamed—thought the fish might be the fabled coelacanth. Hunt had the fish salted on the spot, then had it injected with formalin to preserve its internal organs, and cabled J. L. B. Smith in South Africa.

After some delay, which was laced with confusion and frustration, Smith managed to reach the Comoros and, when he saw the dead fish, he is said to have wept. It was indeed a coelacanth, this time with its organs intact and captured most probably near the creature's actual habitat.

Thereafter, quite a number of coelacanths have been caught for study, first by French researchers and, after the Comoros Islands became independent in the 1970s, by various international groups of scientists. In addition, *Latimeria* has been studied in its natural habitat by direct observations and by videos obtained by submersibles diving in the waters off the Comoros Islands.[2]

A most remarkable chapter in the history of the study of modern

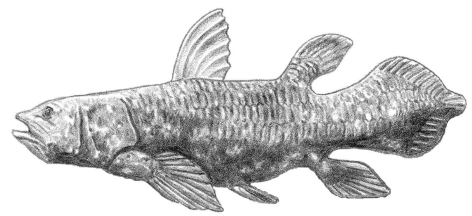

Figure 6.1. The icon of living fossils, *Latimeria chalumnae*

coelacanths began on 30 July 1998. On this date, a population of these lobe-finned fish was discovered by American and Indonesian scientists about 10,000 km east of the Comoros Islands, off the coast of Sulawesi Island in Indonesia. When this coelacanth was first discovered, the only obvious difference between it and *Latimeria chalumnae* from the Comoros Islands was its color. The Comoros coelacanths are renowned for their steel blue color, whereas specimens from the Sulawesi population are brown. Described as a new species, *Latimeria menadoensis*, in 1999 by a host of Indonesians scientists and one Frenchman—L. Pouyaud, S. Wirjoatmodjo, I. Rachmatika, A. Tjakrawidjaja, R. Hadiaty, and W. Hadie—this discovery identifies living coelacanths as more widespread and abundant than previously assumed. Furthermore, it opens an incredible range of ecological, biogeographical, and evolutionary questions associated with these exceptional, albeit still very rare, living fossils.

Latimeria has dual significance in evolutionary biology. First, in the 1930s, extinct coelacanths were thought to be the direct ancestors of the tetrapods (land-living animals, which also include humans), and living coelacanths thus provided a wealth of new information on their nonskeletal anatomy, ecology, and—once living coelacanths had finally been observed in their natural habitat—their behavior. These rare glimpses into their biology have forced evolutionary biologists to reassess the closeness of the relationship between coelacanths and tetrapods, and we now know instead that lungfish hold such a position. Nevertheless, evolutionary biologists remain equally intrigued by *Latimeria chalumnae* through its eminent position as the icon of all living fossils.

What exactly is a living fossil, and why does *Latimeria* qualify? Along with horseshoe crabs, bowfin fish, and opossums, coelacanths are often cited as living fossils, but few scientists have agreed on the precise meaning of the term.[3] Darwin introduced the phrase "living fossils" for forms that are the result of a long survival of lineages and remarkably slow evolutionary rates, both of which he attributed primarily to an absence of ecological competition.[4] Since then, the concept of living fossils often took on an adaptational flavor. For example, Delamare-Deboutteville and Botosancanu envisioned living fossils as organisms that, by virtue of the narrowness of their adaptation, are restricted in their ability to change over time.[5] In contrast, Simpson considered them to be characterized by broad adaptation.[6] Adaptations aside, most assessments of living fossils agree that they share the following: they have survived for a relatively long time—measured in terms of geologic periods—and they exhibit a plethora of primitive characteristics that suggest that they have undergone little evolutionary change over this period of time.

The Transylvanian fauna were what formed the backbone of Nopcsa's ideas in paleobiology and evolutionary theory, among them the connection between body size and habitat area (i.e., dwarfs and islands) and the evolutionary processes that mediated this relationship (endocrine disease and neo-Lamarckian inheritance). It also revealed to Nopcsa that his dinosaurs represented a depauperate assemblage of relicts of a much richer European fauna from earlier Cretaceous times. Although he never discussed this relictual nature of the Transylvanian dinosaur fauna using Darwin's term, living fossil, in what follows, we are going to claim that—taken from the perspective of 70 million years ago—at least one member of the Transylvanian fauna can be regarded as a paleontological example of a living fossil. This is not an oxymoron—we really do mean living fossil, in the sense that it is separated by a long interval of time, and is little transformed, from its closest relatives or its potential ancestor. We will also show that other members of this region's fauna changed at much more normal rates.

Evolutionary rates come in two varieties: taxonomic rates and rates of character change. In extinct organisms, taxonomic rates have generally been estimated from the first occurrence of a particular taxon, its temporal duration, and its diversity, and these rates are often expressed using survivorship curves.[7] Rates of character change typically have been evaluated as changes in the size and shape of a feature (e.g., tooth-crown height) over a given time interval, although rates of change among character complexes have also been analyzed.[8]

Here we're going to try something a little different: instead of measuring rates of character change from what has come directly from the fossil record, we're going to calculate these changes on the basis of what we *don't* have. In order to calculate the differences in rates of character change in some of our Transylvanian dinosaurs and to analyze their significance—in other words, to look for the presence of living fossils in the Late Cretaceous—we must also come to terms with their ghosts.

GHOSTS THAT TELL TIME

Our quest to understand the how and why of differences in evolutionary rates among the Transylvanian dinosaurs requires us to estimate their rates of morphological change, and to calculate these rates we need to know not only the time period, but also the historical pattern of common descent. As we have learned in chapter 2, the latter is what cladograms are all about: portraying the closeness of the relationships of different organisms. In so doing, cladograms also provide information about the relative sequence of evolutionary events—the mutual divergence of new lineages—that produced these organisms. As originally noted by Willi Hennig and later elaborated by Mark Norell, a vertebrate paleontologist and systematist at the American Museum of Natural History in New York City, it is axiomatic from evolutionary and cladistic theory that monophyletic sister taxa must have separated from each other at the same time through the same evolutionary splitting event.[9] It is this historical continuity between paired sister taxa through their common ancestor at splitting events, when combined with temporal information from the stratigraphic record, that allows us to determine the minimal age of splitting and then use it as a "clock" to extend the minimal age of each group beyond the information that comes from stratigraphy alone.

Ghost Lineages and Their Durations

A ghost lineage is that part of an evolutionary tree for which there is no fossil record, but which must have existed because of the continuity through time between all of the ancestors and their descendant.[10] In order to identify the existence of ghost lineages in a clade and thereby calculate their durations, we obviously need to know the clade's phylogeny. That is, we must have as complete, as fully resolved, and as well-tested a cladogram as possible for the group in which we're interested. All of the taxa in the cladogram should be monophyletic, without unresolved putative ancestors. When these aspects of a cladistic analysis are fulfilled, the resulting cladogram is the best we can expect from the most parsimonious treatment of the data. In addition, we must have the best data on the stratigraphic distribution of the various members of this clade. Anything less than all of what is currently available, on as precise a form as possible, will reap as much error in ghost lineage analyses as it would in an improperly constructed cladogram. If these aspects are kept in mind, ghost lineage analysis can meld information from phylogeny and stratigraphy to provide a measure of the quality of the fossil record.

To assess how well a set of first appearances in the fossil record corresponds to the prediction of specific phylogenetic hypotheses, we will use a method that has been called the Sum of Minimum Implied Gaps (SMIG).[11] SMIG analysis pits the first occurrence of a taxon in the fossil record against its relationships, as determined by cladistic analysis, in the following way. Suppose dinosaurs X and Y are each other's closest relatives (figure 6.2). As a result, they shared a common ancestor that is not shared by dinosaur Z or any other life forms. From different sources of information, we know that dinosaur X comes from rocks dated 100 million years ago and dinosaur Y comes from 125-million-year-old rocks. Since ancestors must come before descendants, the ancestor of X and Y has to be at least 125 million years old. We have thus discovered a ghost lineage leading to dinosaur X, amounting to 25 million years of not-yet-sampled history—its ghost lineage duration (GLD). The same follows for the ghost lineage of 5 million years leading from the common ancestor of dinosaur Z and the common ancestor of dinosaurs X and Y. GLDs are obtained by subtracting the earlier age (the earliest stratigraphic occurrence of the form that lacks a ghost lineage) of one taxon from the later age (the earliest stratigraphic occurrence of the form with the ghost lineage) of its sister taxon.

Short GLDs imply that there is not a great deal of missing history in the stratigraphically calibrated cladogram, while long GLDs indicate the

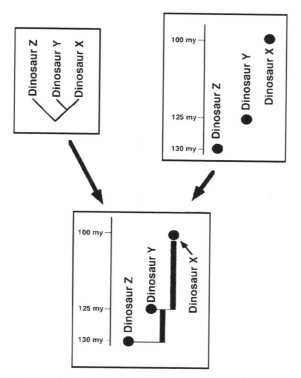

Figure 6.2. Determining ghost lineages and their duration. The diagram in the upper left indicates the phylogenetic relationship of three dinosaur species (dinosaurs X, Y, and Z), while the diagram in the upper right provides the stratigraphic distribution of these same species. Together, the stratigraphic calibration of this dinosaur phylogeny is provided in the lower diagram. Ghost lineages are indicated by shaded rectangles in this lower diagram (see text for a further explanation).

opposite. The ghost lineages for the dinosaurs from Transylvania are of the 10-million-year magnitude, similar to those calculated for other dinosaur taxa. To provide a broader perspective, the ghost lineages for horses and humans are on the order of several million years.[12] This difference between dinosaurian GLDs, on the one hand, and those of horses and humans, on the other, indicates not only that considerable time is missing in dinosaurian history in general, but also that the quality of the fossil record for at least some of the dinosaurs in the Mesozoic is significantly less than that for many of the mammals in the Cenozoic. Do these temporal lapses make it easier to speculate more and be less critical about the history of the Transylvanian fauna? We certainly hope not, and in chapter 7 we will take up this issue of the quality of the fossil record we might expect from a small, isolated region enveloped in the widespread tectonic dynamism that is the Late Cretaceous of eastern Europe.

Anatomical Oddballs and Rates of Evolution

By identifying ghost lineages and estimating their durations, we now have the time part of the equation for calculating rates. Now all we need are the number of characters that change over these intervals. For that information, we return to Courtenay-Latimer's favorite fish, the coelacanth.

Latimeria is considered to be a living fossil not only because of its extensive removal in time from its last known relatives in the Cretaceous, but also because it bears a very close resemblance to those ancient fish. That is, it hasn't changed very much over the course of its history. On the other hand, other living creatures (say, teleost fish) with an ancestry coeval with that of *Latimeria* have changed far more dramatically, both in terms of their taxonomic diversity and their morphology, than have coelacanths—the evolutionary rates of the telosts are profoundly greater than that of *Latimeria*. This difference between the coelacanth and the teleost conditions is one that involves rates of morphological change, a subject that we now want to explore, from the perspective of coelacanths and—naturally—the dinosaurs of Transylvania.

The rates of morphological evolution through geologic time, and, coincidentally, about coelacanth evolution, have recently been investigated by Richard Cloutier, a phylogenetic systematist in the Laboratoire de Biologie Évolutive at the Université du Québec à Rimouski, Québec, Canada. Because these rates are based on history, Cloutier began his analysis with the construction of a cladogram for actinistian fish (the group to which coelacanths belong), providing the shape of the evolutionary tree but not the timing of the branching events (figure 6.3a). Cloutier then calibrated his cladogram with the stratigraphic distribution of the actinistians he was analyzing. For example, because the closest relative of *Latimeria* is a coelacanth known as *Macropoma* from the Cretaceous of Europe, the duration of the ghost lineage leading to the former is approximately 80 million years (figure 6.3b). Based on his

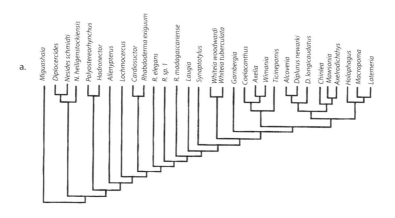

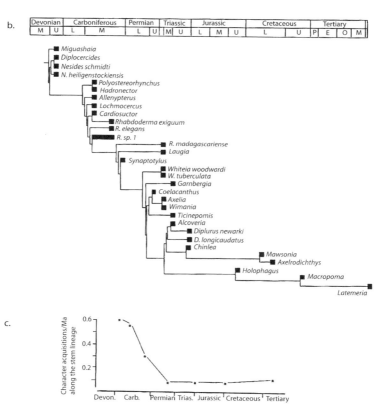

Figure 6.3. (a) A cladogram of actinistian fish; (b) ghost lineages determined from the actinistian cladogram and the stratigraphic distribution of fossil actinistians; and (c) the rate of morphological change, calculated as the number of character changes for each ghost lineage divided by the ghost lineage duration. (After Cloutier 1991)

cladistic analysis, Cloutier had, at hand, all the character changes that took place along each branch of the tree (figure 6.3c). In fact, there were 101 changes distributed among the 17 branching events leading from the bottom of the tree to present-day *Latimeria*. Cloutier was then able to estimate the rates of morphological change through geologic time for these fish. We will take the same approach as Cloutier, using phylogeny and stratigraphy to estimate evolutionary rates for some of the Transylvanian dinosaurs. In so doing, we will address another of Nopcsa's claims about Transylvanian dinosaurs, namely, their supposed primitive nature and their rates of character evolution.

THE GHOSTS OF NOPCSA AND OF TRANSYLVANIAN DINOSAURS

"The Haţeg fauna turns [out] to be nothing else than the poor remains of an older and richer but less known fauna. Thus we have not only to deal with a primitive fauna, but with one in which the genera are reduced in number whereas individuals are abundant."[13] With these words, Nopcsa tied not only dwarfing, but also aspects of character evolution, to his emerging view of the Haţeg region as a Late Cretaceous island. When he identified the Haţeg assemblage as a relict fauna of dwarfs, Nopcsa was arguing that Europe was moving from a more continental to a more insular geography during the Cretaceous, complete with increases in extinction and a slowing down of evolutionary rates. He would also have argued that the further back in time we go, there would have been more habitable area in Europe, and therefore a greater diversity of the kinds of dinosaurs, than we met with in the later Transylvanian fauna. As we will discuss in some detail in chapter 7, the geographic distributions of the closest relatives of the Transylvanian dinosaurs are most often not European, strongly suggesting that we should look elsewhere for their ancestral regions of origin. Now, however, it is important to address the rates of character evolution that Nopcsa used to characterize the Haţeg dinosaurs as relicts.

We turn first to the Transylvanian hadrosaurid *Telmatosaurus* (chapter 2). The sole species, *T. transsylvanicus*, is well diagnosed, recognized by a suite of characters that are not found in any other dinosaur. In addition, it is positioned basally among Hadrosauridae, prior to the split of the latter into hadrosaurines and lambeosaurines.[14] This phylogeny of

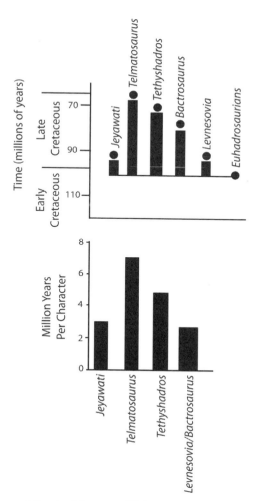

Figure 6.4. A diagram of the ghost lineages of *Telmatosaurus* and its close relatives, Euhadrosauria, *Levnesovia*, *Bactrosaurus*, *Tethyshadros*, and *Jeyawati* (*above*); and the rates of character changes corresponding to these ghost lineages (*below*)

higher iguanodontian ornithopods, when calibrated against stratigraphy, indicates that there is a long ghost lineage between *Telmatosaurus* and the minimal age of its common ancestor with the remaining hadrosaurids, a GLD that is roughly 35 million years long (figure 6.4).[15]

GLD calculations for the two species of *Zalmoxes* follow the same approach (figure 6.5). Because *Z. robustus* and *Z. shqiperorum* are found

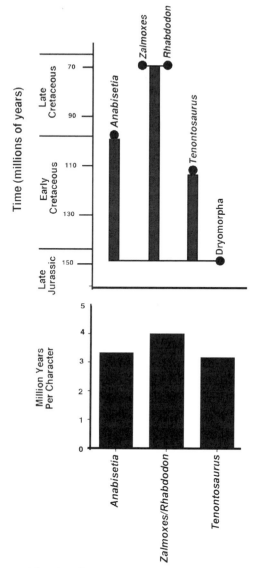

Figure 6.5. A diagram of the ghost lineages of *Zalmoxes* and *Rhabdodon* and their close relatives, Dryomorpha, *Tenontosaurus*, and *Anabisetia* (*above*); and the rates of character changes corresponding to these ghost lineages (*below*)

in the same Sânpetru and Densuş-Ciula beds of Transylvania, they have a coeval stratigraphic distribution, and no ghost lineages can be identified between the two. The same applies to *Rhabdodon* and *Zalmoxes*—both are known from the Campanian and Maastrichtian, and therefore only a short ghost lineage can be identified here, perhaps upward of a few million years. However, when these two ornithopods are tethered to Iguanodontia (their immediate sister group), we get our first glimpse of a ghost lineage leading to Rhabdodontidae. The earliest known iguanodontians (*Dryosaurus* and *Camptosaurus* from the Late Jurassic) are older than both *Rhabdodon* and *Zalmoxes*, so the ghost lineage leads from the common ancestor of all involved to that of *Zalmoxes* and *Rhabdodon*. With stratigraphic calibration, this amounts to a GLD of more than 80 million years.

These GLDs clearly indicate that a good deal of the tangible history of these lineages is thus far invisible to us. Such a pity, this gob of spit in the face of every paleontologist trying to make sense of the evolutionary significance of fossils (paraphrasing Henry V. Miller in *Tropic of Cancer*[16]). Bothersome though they may be, long ghost lineages could be all we're ever going to get, because they're the most likely expected consequence of the ephemeral terrestrial habitats that existed during the Cretaceous in Europe (chapter 4). The patchiness of available land in time and space drastically decreases the probability that nonmarine sediments will contribute to the geologic and fossil records, which in turn reduces our ability to sample many of the Transylvanian lineages. In other words, it's our previous bugbear of European isolation in the Cretaceous again, where terrestrial habitats must have been rare, at best, over the long run. GLDs may shorten as we discover new taxa whose phylogenetic position fits within these long ghost lineages, but these improvements will probably be hard won for places such as Europe in the Cretaceous.

Such a long ghost lineage might suggest that *Telmatosaurus* was a Late Cretaceous living fossil—its latest Late Cretaceous occurrence (approximately 68 million years ago) stands in contrast with its closest phylogenetic relationships, dating back to the late Early Cretaceous (some 103 million years ago). But long expanses of time do not make a living fossil. What about the rate of evolutionary change over this ghost lineage? We can calculate the minimum rate of character changes accumulated along ghost lineages from *Telmatosaurus* to its most recent common ancestor with the remaining hadrosaurids. These transformations amount to those features that mark a species as being unique. Returning to figure 6.4, *Telmatosaurus* is currently diagnosed by five characters, which had to evolve during the 35 million years of its ghost lineage.[17] That's one character every 7 million years, a very slow evolutionary rate indeed. A number of closely related iguanodontians evolved at much higher rates. For four of these taxa, GLDs range from 5.4 to 28.5 million years, during which approximately two to six character transformations took place. That amounts to approximately three characters per million years for the *Levnesovia/Bactrosaurus* clade. Interestingly, the number of characters per million years for *Tethyshadros*, the other taxon from the Late Cretaceous of Europe, is higher (4.75 characters per million years), between that of the *Levnesovia/Bactrosaurus* clade and *Telmatosaurus*.

Zalmoxes is another dinosaur from Transylvania for which there is good phylogenetic and stratigraphic information. As we learned earlier, there are two species of *Zalmoxes* in Transylvania, the more common *Z. robustus* and the rarer *Z. shqiperorum*. Figure 6.5 shows that these two species are most closely related to *Rhabdodon priscus*, as members of Rhabdodontidae. Rhabdodontidae is then closely related to the great clade of ornithopods called Iguanodontia, which includes *Tenontosaurus*, *Dryosaurus*, *Camptosaurus*, and *Iguanodon*, as well as *Telmatosaurus* and the remaining hadrosaurids (chapter 2). Thereafter, more primitive relationships are with *Anabisetia* and a clade consisting of *Thescelosaurus* and *Parksosaurus*.

Ghost lineages for this suite of dinosaurs were determined in the same way as we did for *Telmatosaurus*. With a GLD of more than 80 million years, rhabdodontids are separated by approximately 20 characters from their common ancestor with Iguanodontia. That's one character every 4 million years, slightly less than twice the rate of character evolution in the *Telmatosaurus* lineage.

How do these character change rates stack up against those of other dinosaurs? In answer, we've tracked the number of character changes for the ghost lineages of several other, closely related taxa within Ornithopoda in a similar fashion. For example, the ghost lineage of *Tenontosaurus* (Early Cretaceous of the western interior of the United States) comes out to one character every 3 million years. Other rates within Ornithopoda generally fall out around 1 million years for each character change, although for the Argentinean *Anabisetia* the figure is more like 3.25 million years (what this elevated number means is unclear—is it because these ornithopods are themselves living fossils, or because there are lots more of these kinds of dinosaurs yet to be discovered in South America?). Other dinosaur groups for which these estimates have been made show a similar pattern.[18] For thyreophorans (remember, this group includes, among other taxa, stegosaurs and ankylosaurs like our own *Struthiosaurus*), ceratopsians, sauropods, and nonavian theropods, on average it takes slightly over 3 to about 4.25 million years for a character to change. Interestingly, for avian theropods the rate climbs to one character change per 200,000 years. It is difficult to say whether this shift is real, reflecting a change in the life-history strategies and evolutionary dynamics of these small theropods, or if, instead, it is a product of the intense work that has gone into understanding their phylogenetic relationships. In either case, future discoveries are bound to make this an exciting realm of dinosaur evolutionary biology.

Returning to our treatment of the Transylvanian dinosaurs, the rate of evolution in the lineage leading to *Zalmoxes* is similar to that found both in its immediate clade and in other dinosaurs elsewhere in the world; whereas it is much slower—about the same rate as the living fossil, *Latimeria*[19]—in the lineage leading to *Telmatosaurus*. Without good phylogenetic analyses for titanosaurid sauropods, nodosaurid ankylosaurs, and dromaeosaurid theropods, it's not yet possible to calculate the evolutionary rates of *Magyarosaurus*, *Struthiosaurus*, and the Romanian raptor.

Even without rates for these remaining Transylvanian taxa, what we know from *Zalmoxes* and *Telmatosaurus* suggests that Nopcsa's faunistic view of the Haţeg assemblage—as primitive and impoverished by an ever-dwindling habitable "Europe"—is probably incorrect. Even though we have sampled only two taxa, their evolutionary rates, determined through phylogeny and stratigraphy, are definitely far from uniform and slow: character transformations leading to the hadrosaurid *Telmatosaurus* creep as laggardly as those leading to *Latimeria*, whereas their pace is much more normal in the ghost lineage leading to *Zalmoxes*. Late Cretaceous life in Transylvania, and the evolution that preceded it, was

Living Fossils and Their Ghosts   161

much more complicated—and richer for it—than Nopcsa had originally imagined.

We've now moved our understanding of the Transylvanian dinosaurs a step farther forward. In some (if not all) cases, they were neither the faunistic residue nor the uniformly phylogenetically arrested descendants of earlier European biotas. Each, as might be expected, arose independently, and they happened to find themselves living together and being preserved together in the uppermost Cretaceous rocks of Transylvania. Some dinosaur species had evolved a great deal relative to their ancestors, whereas others hardly evolved at all. As with *Latimeria*—the icon for all living fossils, with its restricted deepwater habitats off the coasts of the Comoros Islands in the western Indian Ocean and Sulawesi Island in the South Pacific—the isolation of the Transylvanian dinosaurs produced at least one Late Cretaceous living fossil—the slowly evolving *Telmatosaurus*. At the same time, *Zalmoxes* took another road, one that involved more normal rates of evolution. Nonetheless, many more questions remain to be asked than we have answered thus far during our exploration of evolutionary rates. For example, what drove the differences in the evolutionary rates of character change among the Transylvanian dinosaurs? From what we determined earlier in this chapter, these differences are not merely an effect of respective ghost lineages, but may instead be a function of the timing of a creature's arrival in the Transylvanian region. If it is historical biogeography and not ghost lineages that are relevant here, then we should look to regional changes in European terrestrial environments, and also to what was happening on other continental landmasses. In addition, we should ask why the isolated Transylvanian region included a primitive member of one clade and not of others. Depending on the connections enabling colonization of the Transylvanian region over time, this area may have been available for early or late members of contemporary clades distributed elsewhere. Or maybe there is some sort of ecological effect, in terms of how resources or local habitats were available to and used by the Transylvanian dinosaurs. In our next chapter, we turn to these issues of historical biogeography and colonization, their relationships to body size and faunistics, and, ultimately, to the interplay of contingency and selection as they are played out in evolutionary history.

CHAPTER 7

Transylvania, the Land of Contingency

You're standing at one of those classic 1970s pinball machines—no doubt the "Bally 4 Million BC," in keeping with our paleontological theme—and you pull back the plunger and release the ball on its initial trajectory. As the silver projectile cavorts with the zap and ding of the bumpers and the slap of the paddles, the score mounts. Eventually, the game ends when the ball drops into the machine and you cram another coin into the slot as you obsessively try to beat your previous score. Game after game, the itinerary of the pinball is different, guided or constrained, to be sure, by the rails, the paddles, and the bumpers, but the cascade of small variations in speed, angle of impact, or body jostles make every game unique. This improvisational dance goes to the heart of pinball's appeal. Myriad small factors—such as scratches on the ball, the humidity of the machine, dust on the plunger, and the twist of the wrist—affect each and every pathway, even before the ball is unleashed. This unpredictability is the essence of historical contingency.

As much as it is a recipe for making pinball a challenge and enjoyment, historical contingency is also what holds our attention in the arts, especially in film. For example, Frank Capra's 1946 film, *It's a Wonderful Life*,[1] has been invoked by Stephen J. Gould, the late Harvard professor, evolutionary theorist, and paleontologist, as a present-day metaphor for the role of chance in human history.[2] In this Christmas favorite, George Bailey, depressed and suicidal over his inconsequential existence, is shown by an angel what his community would have looked like if it hadn't been for all his good deeds over the years. In providing him with an alternative "what-if" version of his existence, the angel convinces George to return to his family and his way of life. Countless other films look at the other side of life's singularity by exploring how seemingly insignificant events can unpredictably produce cataclysmic results. Of these, we especially like Terry Gilliam's 1985 Orwellian extravaganza, *Brazil*. This film is one long nightmare of terrorist bombings, late-night shopping, true love, creative plumbing, and brainwashing, ultimately based on the intersection of a dead beetle and a typing machine.[3] Instead of George Bailey, we have Sam Lowry, a harried civil servant in a convoluted and inefficient society, who escapes to dreams of flying, overpowering enemies, and spending eternity with the woman in his fantasies. At the same time, it is Gilliam's colossal joke about chance and the unpredictability of the future. The small and apparently insignificant changes that are the lynchpins of both of these films become the catalysts for cascades of historical differences. *It's a Wonderful Life* shows how a mere person can alter historical pathways, whereas *Brazil* does so by linking an unpredictable triviality to the overthrow of a man's life. Change the initial conditions even slightly, and without apparent importance, and the path of history can be radically altered.

In coming to this place in our book, we now will reach past the dinosaurs of Transylvania to explore whether unpredictable events can have a uniquely important influence on the pattern of evolutionary history; we are going beyond the "who," "what," and "where" of the Transylvanian fauna to a framework that will allow us to explore "how" and perhaps even "why" questions. How did the creatures come to inhabit this part of eastern Europe? Who are the island immigrants, and are they a random sample of mainland faunas? If they indeed were dwarfs, how and why might they have become so?

Historical contingency—the unpredictability of life's history from particular, often small-scale events—will take a prominent place as we build a foundation to answer these and other questions about the dinosaurs of Transylvania. Even one of the fundamentals of evolutionary processes—speciation—is fraught with its fair share of unpredictability. Speciation is the formation, through genetic divergence, of two or more descendant species from one ancestral species. This process is the motor for the diversification of life, and, when rendered in terms of cladogenesis (i.e., the generation of branches in a cladogram; see the discussion of cladistics in chapter 2), it forms the backbone for identifying the overall pattern of evolutionary history.[4]

Speciation is not just enhanced by, but, more importantly, it is dependent on a plethora of unpredictable conditions, particularly due to insularity. Contingency should be more common, and have greater effects, in situations of isolation; insularity should showcase historical contingency. What is crucial is reproductive isolation. When gene flow within a once-contiguous, interbreeding population is sufficiently interrupted, genetic divergence takes place between two or more portions of that original population. This divergence can occur in the same place as the ancestral population (sympatry), along the length of a geographically contiguous population (parapatry), or in geographic isolation from the ancestral population (allopatry).[5] The evidence is extensive for allopatric speciation, which—despite some objections[6]—is generally regarded as the most influential theory of speciation. In addition to its prominence in evolutionary theory, we will be discussing it here in order to scrutinize the relationship of the Transylvanian fauna to geographic isolation.

The basis for allopatric speciation is the development of a geographic barrier separating an ancestral population into two subsequent descendant populations—think of a new mountain range or a river cutting through a formerly continuous population. If such barriers could split a species for a long enough period of time, gene flow between the two groups would cease and each would then evolve separately. Depending on the specifics, mutation, genetic drift, and/or natural selection would alter the genome of the two descendant populations over time. If these populations rejoined after they had diverged sufficiently, it is unlikely that they would successfully interbreed, either because mating would yield unviable or sterile offspring, or they might not recognize each other as members of the same species and hence would not interbreed. When the former applies, we speak of postmating or postzygotic isolating mechanisms, and, for the latter, premating or prezygotic isolating mechanisms. In either case, geographic isolation has led to reproductive isolation and thereby to speciation.

Preeminent among the cases of allopatric speciation are those involving peripherally isolated populations. Ernst Mayr, one of the twentieth century's most influential evolutionary biologists, originally described the basis for speciation through peripheral isolation—sometimes called peripatric speciation—and argued that it was one of the most effective means and commonest way of producing two species where once there was one.[7] Peripatric speciation involves the founding of a small second population—a few original individuals, or perhaps a single fertilized female in the most extreme case—through geographic isolation. The colonization of an island by a small population is the most cited example of what has been termed the founder effect.[8] Should this group remain geographically separated from the main population for a sufficient period of time, they, too, will undergo the same divergence we've just discussed generally for allopatric speciation.

An especially important feature of peripatric speciation is that it would proceed very quickly. First, the founding population will carry only a small sample of the genetic reservoir of the ancestral population. Not only is there such random genetic loss, but a rare gene in the main population may also start being passed on with relatively high frequency in the founder population.[9] These unpredictable sampling errors in gene frequencies—genetic drift—will be especially strong determinants of the subsequent genomes, so long as the population size remains small. Even though the chance of survival of this small peripheral population may be low,[10] it is certainly not zero, and when the founder population increases to a less vulnerable size, natural selection will begin to take over, reflecting conditions that could be quite different from those of the more widespread population. In this way, the speciating population rapidly passes from one well-integrated and stable condition, through a highly unstable period, to another period of balanced integration.

There's certainly a lot of dumb luck to peripatric speciation. Peripheral populations may be well adapted to their particular conditions, but, at the same time, their geographic separation from the main population is an unpredictable event. So are the details of genetic loss during separation, with the founders carrying a small, unrepresentative portion of the genome of the ancestral population. Chance also comes into play while the population remains small—gene frequencies are free to drift up and down randomly, without the controlling hand of natural selection. In this way, historical contingency is not just a possibility, but instead is a necessity in producing rapid speciation through isolation of the kind thought to exist in nature.

CONTINGENCY AND THE TRANSYLVANIAN DINOSAURS

If anywhere in the world should reveal the power of contingency, it would be somewhere like Transylvania in the Cretaceous, where historical possibilities and pathways can be altered by chance events in the small, insular populations living there. Because of the geographic isolation of the place, coupled with its considerable tectonic flux (chapter 3), we should find an unusual agglomeration of animals and ecological relationships in Transylvania when compared with those more typical of the nearby North American and Asian landmasses, where the coastal plains were much more extensive, luxuriant, and tectonically quiet. Just who arrived in the Transylvanian region, and from where, are both far from predictable. Equally serendipitous are differences in the structure of predator-prey relationships, patterns of growth and development, intraspecific life histories, social structures, and evolutionary dynamics. When might these colonizers have arrived? Who interacted with whom? What changes in their features might have occurred? These are some of the questions we'll be pursuing in our attempt to assess contingency in the history of the dinosaurs from Transylvania. Whatever answers we can bring to these questions, they must certainly be couched in the one thing we learned about the Transylvanian fauna in this and the previous chapter—its susceptibility to the vagaries of unpredictable evolutionary events due to its isolation.

BACK TO THE PAST

In a shallow sea, a parcel of land becomes available for whatever terrestrial organisms come its way. Yet who's to say which organisms? The issue of who successfully colonizes a new piece of ground is complicated by the size of the property; its proximity to other terrestrial regions; environmental differences in temperature, rainfall, and so forth; a species' potential for transoceanic migration; which species populate neighboring regions; which species might already have arrived on the newly available land; and, above all, *chance*.

In this chapter, we will look at Transylvania as more than a place where dinosaurs once lived some 70 million years ago. With different eyes than we had at the beginning of this book, we now see more than a jumble of fossils, the remains of which have long been collected in the picturesque northern foothills of the Retezat Mountains of western Romania. First, by virtue of both the studies conducted by Franz Nopcsa and subsequent research efforts, we've gone well beyond merely recognizing various creatures from a collection of scattered bones found in western Romania: the duck-billed *Telmatosaurus*, the solidly built *Zalmoxes*, the long-necked *Magyarosaurus*, the armored *Struthiosaurus*, and the thus-far imperfectly understood predatory dinosaurs. Through paleobiological inference, these dinosaurs have become more than just entries in a Transylvanian faunal list. Second, we've used cladistics to put these Late Cretaceous denizens into their community of descent—siblings and cousins, as it were—with dinosaurs elsewhere in the world. Third, we've see them in as much of their complex terrestrial habitat as data and conjecture allow us to reproduce. Fourth, the small size of these dinosaurs—initially noted by Nopcsa—is now given a heterochronic context. Fifth, we've established that Transylvania was a special haven that existed in relative isolation from other terrestrial habitats at the end of the Cretaceous.

This understanding of the Transylvanian dinosaurs as not only once living, but also evolving creatures has provided us with the basis for exploring themes introduced earlier in this book: the paleogeographic relationships of the Cretaceous landforms of Europe, colonization and faunal balance in isolated environments, and the nature of body-size changes and life-history consequences. We are now at the logical place to add historical biogeography to this soup. We have given reasons for expecting that historical pathways should reflect the influence of chance events within the broad mediation of the laws of nature, rather than merely straightforwardly following the ever-present operation of these laws. Such unpredictability gives organisms the opportunity to circumvent the status quo—and the risks of extinction—and thereby expand their own evolutionary opportunities and innovations. Consequently, we will go out of our way in what follows to emphasize the theme of historical contingency, not only when interpreting the biogeographic history of the Transylvanian dinosaurs, but also as we integrate this information with their paleoecology, changing body sizes, life-history strategies, and phylogeny. Finally, we'll let our Transylvanian dinosaurs wander on into the realm of evolutionary theory, in particular the role of chance in the regulation of diversity—the stuff of what is known as the Red Queen hypothesis.

OF TRAVELING DINOSAURS AND MOVING CONTINENTS

Like organisms everywhere, the Transylvanian dinosaurs were the product of their individual histories. Obviously, one aspect of these histories is their arrival in the region of what is now western Romania. Nopcsa's approach to where these dinosaurs (and the remainder of the Transylvanian fauna) originated was to look solely within Europe. For example, he compared his Transylvanian hadrosaurid with *Iguanodon* from the rich faunas of the Early Cretaceous Wealden faunas of England, Belgium, and France,[11] and why not—both *Telmatosaurus* and *Iguanodon*, although separated by 50 million years, were European members of Ornithopoda. Likewise, Nopcsa directly compared the other members of the Transylvanian fauna with their European relatives from the Early Cretaceous as he attempted to understand how his peculiar dinosaurs arose. We agree that Nopcsa's was a good and logical beginning, but why restrict ourselves to Europe? Why not expand the comparisons to all ornithopods throughout the world and evaluate their areas of origin using their phylogenetic relationships?

In order to build a global biogeographic history of the clades containing our Transylvanian dinosaurs, we will need to know not only about *Iguanodon*, *Hypsilophodon*, *Hylaeosaurus*, and *Pelorosaurus* from a Europe of earlier times, but a great many more dinosaurs from elsewhere. Now we've got to deal with the likes of *Probactrosaurus*, *Gobisaurus*, and *Opisthocoelicaudia* from Asia; *Anabisetia*, *Gasparinisaura*, and *Saltasaurus* from South America; *Muttaburrasaurus*, *Leaellynasaura*, *Minmi*, and *Austrosaurus* from Australia; *Ouranosaurus*, *Valdosaurus*, and *Malawisaurus* from Africa; and *Tenontosaurus*, *Thescelosaurus*, *Pawpawsaurus*, and *Alamosaurus* from North America. Bringing these additional taxa into our investigation, however desirable, should also give us pause, because of the uncertainties arising from the effects of patchy geographic coverage (now increased to worldwide scales) and additional aspects of stratigraphic incompleteness. What we don't know is still more than we know, so any attempt to infer biogeographic patterns may be condemned to failure before we even begin. Nevertheless, we don't intend to give up here without making an effort to reconstruct a history based on what we do know. In so doing, we turn again to the phylogenetic relationships of our Transylvanian dinosaurs, this time combining cladistic analysis with the global geographic occurrences of the groups within which they fall. This can be done by combining geographic data with phylogeny. If this approach sounds familiar, it should—we are again going to overlay *a posteriori* optimization of the geographic occurrences of the Transylvania taxa and their close relatives onto their phylogeny in order to reconstruct their biogeographic history (chapter 5). In this search for the source areas where particular dinosaurs evolved, we will rely not only on the paleogeographic reconstructions of this part of Europe through the Cretaceous (chapter 4), but also on any other landforms brought into consideration by the relationships of the taxa involved. This certainly means we will also have to take into account the paleogeographic conditions and proximity of parts of Asia, South America, Africa, and Australia.

Before we begin looking at the biogeographic history of Transylvania from a worldwide perspective, a word of caution is necessary. Inferences about global biogeographic histories can only be as good as the fossil record will allow, and the biases inherent in this record should be admitted from the start. Ours has to do with the skewed geographic sampling of dinosaurs throughout the world. For example, we know that North America, Asia, and Europe each contain about 28% of the number of world-wide locations during the Early Cretaceous, whereas the remaining continents—South America, Africa, and Australia—each contribute an average of 5% (not surprisingly, Antarctica has thus far contributed nothing). In the Late Cretaceous, North America dominates at 31%, followed by Asia at 24%, Europe at 18%, and South America at 13%. Africa, Australia, and Antarctica together total a measly 6%.[12] Even though our knowledge of dinosaur distribution around the world is always on the rise (as witnessed by the abundance of media accounts), it is also dominated by the big three—North America, Europe, and Asia. Consequently, any search for the area of origin for any of the Transylvanian dinosaurs and their immediate clade is biased in favor of the big three (which constitute the supercontinent of Laurasia) and away from the

Gondwanan landmasses, simply because of the number of fossil locations available to us at this point, rather than being based on real biogeographic history. For now, there's nothing we can do about this problem.

With this cautionary aspect to inferring biogeographic history in mind, we turn first to *Telmatosaurus transsylvanicus*. Nopcsa originally thought this dinosaur lay at the base of Hadrosauridae, a position that has been confirmed over the ensuing years, both through the discovery of many new hadrosaurids and other iguanodontians, and by the application of cladistic methods to understand the general shape of the group's family tree (chapter 2). We've already used this information, plus stratigraphic data, to identify hadrosaurid ghost lineages and their duration while examining rates of character change (chapter 6).[13] What might this approach additionally tell us about the source area for the lineage leading to *Telmatosaurus* and its arrival in what is now Transylvania?

The closest relatives of *Telmatosaurus*—among them, euhadrosaurians, *Bactrosaurus*, and *Levnesovia* (chapter 2)—probably originated in Asia at some time during the latest Early Cretaceous (roughly 103 million years ago).[14] When all of these locations are plotted on a cladogram (figure 7.1), it is unclear where the source area of Hadrosauridae was. It could have been either Asia or North America; many present-day studies support an Asian origin for Hadrosauridae.[15] *Telmatosaurus* and *Tethyshadros* (or their ancestor) can therefore be interpreted as having dispersed on their own to Europe. Similarly, forms such as *Jeyawati* and the small clade of *Eolambia caroljonesa* and *Protohadros byrdi*,[16] the former from the early Late Cretaceous and the latter two from the earliest Late Cretaceous, represent independent migrations to North America from Asia.

We'd like to conduct a little thought experiment concerning the area of origin for hadrosaurids. In doing so, we will push the biogeographic distribution data as far as they allow, hoping that what they indicate about the area of origin for individual clades will not be outweighed by information about them not yet known to us, because it is hidden in their ghost lineages. If that is the case, we can then only look to future discoveries that sometimes, despite the odds, produce better phylogenetic resolution for whatever questions are under consideration (here, biogeography). Expanding the data envelope involves our attempt to resolve the

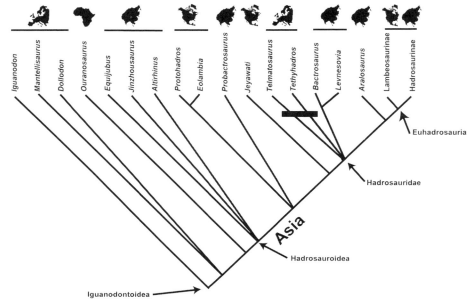

Figure 7.1. A simplified cladogram of higher ornithopods that also includes geographic information, indicating Asia (or possibly North America) as the area of origin for Hadrosauridae and the dispersal of *Telmatosaurus* to Europe, as shown by the bar

previously unresolved relationships of *Telmatosaurus*, *Tethyshadros*, the clade of *Bactrosaurus* and *Levnesovia*, and Euhadrosauria, and looking to Europe for other close relatives.

Tethyshadros insularis: Only a few years ago, the prospect of Europe being the source area of any dinosaur clade was poor to nonexistent. Nevertheless, in the vicinity of Villaggio del Pescatore (along the Gulf of Trieste in northeastern Italy, near its border with Slovenia), a few fragmentary dinosaur bones were discovered in the 1980s, found in the eroding walls in an old abandoned limestone pit.[17] Excavation of the site began in 1992, and by the end of the century this quarry, whose limestones date to the late Santonian (approximately 84 million years ago), had yielded a nearly complete skeleton (figure 7.2a) and the remains of three other individuals, all pointing to a small (4 m long) iguanodontian. These wonderful specimens were studied in detail in 2009 by the Italian paleontologist Fabio Dalla Vecchia,[18] who noted that this dinosaur has many features uniting it with Hadrosauridae,[19] lacks critical characters

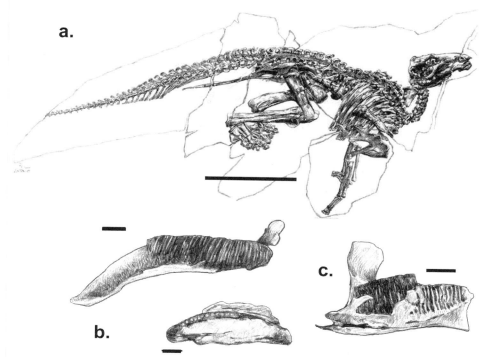

Figure 7.2. (a) *Tethyshadros insularis*; (b) a maxilla and dentary referred to *Pararhabdodon isonensis*; and (c) a dentary of the Fontllonga hadrosaurid. Scale = 100 cm (a), 5 cm (b, c). ([a] after Dal Sasso 2003; [b] after Casanovas et al. 1999a; [c] after Casanovas et al. 1999b)

that unite Euhadrosauria, and has several features of the skull, dentition, hand, and pelvis indicating that it is not *Telmatosaurus*.

Pararhabdodon isonensis: The wooded mountains of northeastern Spain, which rise to their greatest height as the Pyrenees, shelter the region of Catalonia, the original home of Salvador Dalí. There, in central Lleida Province, halfway between Barcelona on the Mediterranean and the small, landlocked country of Andorra, is one of the most prolific areas for the discovery of dinosaur bones, footprints, and egg nests.[20] Our particular interest in the dinosaurs from Lleida is focused on the village of Isona, for here, in 1985, the remains of *P. isonensis* were recovered from rocks of Maastrichtian age, like those of Transylvania.[21] It was originally described by María Lourdes Casanovas-Cladellas and José Vicente Santafé-Llopis (Institut de Paleontologia "M. Crusafont," Sabadell), and

Albert Isidro-Llorens (Institut Guttman, Barcelona). Based on postcranial elements, this new species was thought to be a close relative of an individual estimated to be about 6 m in length (figure 7.2b).[22] Most recently, it has been reexamined by Albert Prieto-Márquez and Jonathan Wagner.[23]

Pararhabdodon clearly is a member of Hadrosauridae (the dentary, unfortunately preserved without teeth, indicates the presence of small teeth arranged in a dental battery, and the deltopectoral crest [not illustrated] is large and angular in profile). Additional material of *Pararhabdodon* has been described from the Maastrichtian in the Upper Aude Valley of southern France, some 150 km to the northeast, over the Pyrenees, from Isona. Both the Spanish and the French specimens indicate that *Pararhabdodon* is not *Telmatosaurus* or *Tethyshadros*. It is either a non-euhadrosaurian hadrosaurid[24] or a very basal lambeosaurine.[25]

The Fontllonga jaw: A second specimen from Spain, thus far consisting of a small dentary (figure 7.2c), was also recently discovered in Lleida Province, in beds contemporary with those of Isona, but 25 km to the south. Known as the Fontllonga hadrosaurid,[26] this form cannot be referred to either *Telmatosaurus, Tethyshadros,* or *Pararhabdodon.* Its small teeth, organized into a dental battery that continues all the way to the front of the jaws, bear an asymmetrically placed median ridge. These features indicate that the Fontllonga hadrosaurid fits between *Telmatosaurus* and Euhadrosauria.[27]

So, by our count, that's three basal hadrosaurids and near-hadrosaurids from Romania, Italy, Spain, France, and European Russia. Given the tree topology resolving *Telmatosaurus, Tethyshadros,* an Asian clade (*Levnesovia* and *Bactrosaurus*), and Euhadrosauria, we now have before us the possibility that Europe was the source area for all of Hadrosauridae (figure 7.3). This interpretation, of course, depends on the resolution of these new European forms, but if the phylogenetic interpretation we've outlined here is more or less correct, it has the effect of shifting hadrosaurid origins in the Early Cretaceous from the broad coastal plains of Asia or North America in the Early Cretaceous to the much smaller, more isolated, and therefore much more evolutionarily volatile landmasses of Europe.

In order to examine the other dinosaurs from Transylvania in the same way, we need to travel to southern France, northern Spain, and

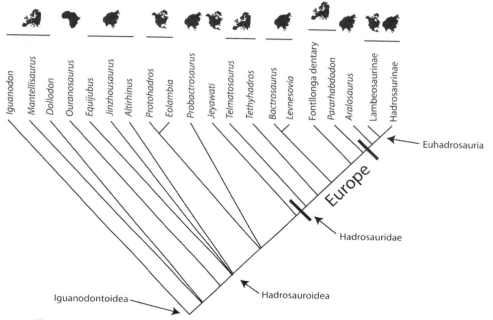

Figure 7.3. A simplified cladogram of higher ornithopods that includes a tentative placement of new hadrosauroid discoveries, indicating the area of origin of Hadrosauridae in Europe and the dispersal of remaining hadrosaurids to North America and Asia

westcentral Hungary. The region extending from the sprawling vineyards of Provence, along the Mediterranean coast in the east, to the green meadows and deep forests of the valley of the Garonne and the foothills of the Pyrenees, in the west, has produced one of the best Late Cretaceous records of dinosaurs in all of Europe (chapter 4).[28] Numerous taxa—including possible dromaeosaurid and avialan theropods, nodosaurids, ornithopods (*Rhabdodon*, *Pararhabdodon*, and thus-far indeterminate hadrosaurids), and titanosaurian sauropods—have been collected from the approximately 20 localities in this region. These faunas of southern France are presently under study by Jean Le Loeuff at the Musée des Dinosaures in Espéraza and Eric Buffetaut at the Laboratoire de Géologie, École Normale Supérieure, in Paris.

In Spain, Laño has one of the most important terrestrial vertebrate assemblages to be discovered in recent years. Located 70 km south of the warm waters of the Bay of Biscay, near the city of Vitoria (and therefore between the European and Iberian tectonic plates), Laño is a small village in the Burgos region of the Basque Country (Euskal Herria in native Euskeran). This site, in an abandoned sand quarry, was discovered in the 1980s and worked by Humberto Astibia and his coworkers from the Euskal Herriko Unibertsitatea in Bilbao, as well as by numerous researchers from Spain and France since then.[29] Over the past 20 years, some 40 species have been recognized, among them new fish, squamates, turtles, crocodilians, dinosaurs, and pterosaurs, all of which are slightly older (late Campanian) than the dinosaur fauna from Transylvania. Thus far, the Laño dinosaurs consist of the euornithopod *Rhabdodon priscus*, some poorly preserved hadrosaurids, indeterminate theropods, and *Lirainosaurus astibiae*, one of the best-known sauropods from Spain.[30]

Finally, in Hungary, north of Lake Balaton, central Europe's largest freshwater lake, this part of the forested Transdanubian highlands, known as the Bakony Mountains, were once home to surface bauxite mining. Iharkút is one of these places. Discovered in 2000 by Attila Ősi and András Torma, this site has now produced fish, amphibians, squamates, crocodilians, pterosaurs, and dinosaurs that are 10–15 million years older than the other Late Cretaceous faunas in Europe. Among the dinosaurs are thus-far indeterminate theropods; abundant material of a nodosaurid ankylosaur called *Hungarosaurus tormai* (figure 7.4a); teeth and a femur indicating a rhabdodontid that appears to be different from both *Zalmoxes* and *Rhabdodon* (figure 7.4b–e); and, most recently, *Bagaceratops kozmai*, a very important and unexpected neoceratopsian. Conspicuous in their absence from Iharkút are sauropods and hadrosaurids.

We turn from these localities first to *Zalmoxes*, *Rhabdodon*, and the Iharkút rhabdodontid. Because all are known only from Europe and nowhere else, we draw the conclusion that this clade is endemic to southern Europe (more precisely, the central part of the northern Neotethyan region) no later than about 80 million years ago. Prior to that, the record of the iguanodontian clade, the closest relative to the clade of *Zalmoxes* and *Rhabdodon*, dates from the Late Jurassic (the age of *Camptosaurus* and *Dryosaurus*; probably near the boundary between the Kimmeridgian and Tithonian, approximately 151 million years ago) of western North America. These dinosaurs indicate that the geographic source area of the

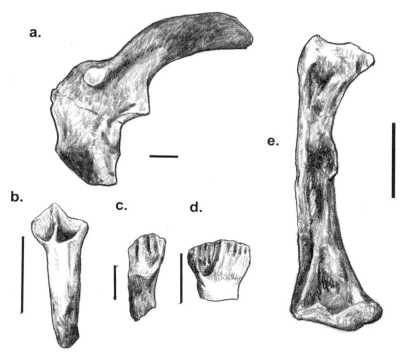

Figure 7.4. (a) A left scapulocoracoid of *Hungarosaurus tormai*; (b) a tooth from a rhabdodontid dentary; (c) a rhabdodontid maxillary tooth; (d) a rhabdodontid maxillary tooth; and (e) a left rhabdodontid femur, in caudal view, all from the Late Cretaceous Iharkút fauna of Hungary. Scale = 5 cm (a), 10 mm (b), 3 mm (c); 6 cm (d); 5 cm (e). ([a] after Ősi 2004; [b–e] after Ősi et al. 2003)

lineage leading to *Zalmoxes*, *Rhabdodon*, and the Iharkút rhabdodontid was North America (figure 7.5).[31]

Likewise, the lineage leading to *Struthiosaurus*, presently known from both Transylvania and Franco-Iberia, dispersed from North America to Europe no later than the Late Jurassic (figure 7.6). The three species of *Struthiosaurus* have, as their successive closest relatives, well-known nodosaurids from places such as Utah, Montana, Wyoming, and Texas in the United States. On the basis of this distribution, it's a good bet that the common ancestor of *Struthiosaurus* hails from North America, and that its migration to Europe resulted in a modest, within-species endemic radiation in the Late Cretaceous. *Hungarosaurus*, the best-known member of the Hungarian Iharkút fauna,[32] probably is a second

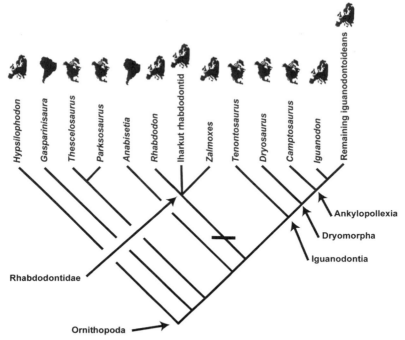

Figure 7.5. A simplified cladogram for basal ornithopods that also includes geographic information, indicating North America as the area of origin of Ornithopoda, and the subsequent dispersal to Europe of the lineage leading to *Zalmoxes* and *Rhabdodon*

migrant from North America to Europe. However, the two may constitute the product of a single dispersal to Europe; we just can't tell at the moment.

From its phylogenetic position and its geographic distribution, *Magyarosaurus* shared Late Cretaceous Europe with at least two other titanosaurs, *Lirainosaurus* from Spain and *Ampelosaurus* from France (figure 7.7). *Ampelosaurus*, the best represented of these European titanosaurs, is now known from abundant isolated material pertaining to nearly the entire skeleton and, more recently, from an articulated skeleton.[33] *Lirainosaurus* is nearly as well known, from the majority of a single skeleton. For a long time, titanosaurs were thought to originate somewhere in the southern continents (Gondwana), but recent phylogenetic studies suggest that the jury is out on their place of origin.[34] Nevertheless, *Magyarosaurus*, *Ampelosaurus*, and *Lirainosaurus* seem

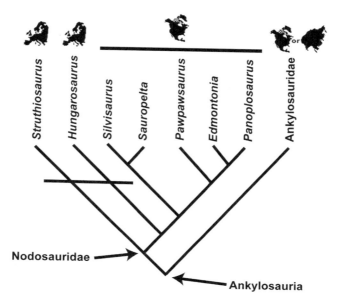

Figure 7.6. A simplified cladogram for the nodosaurid ankylosaurs that also includes the dispersal of *Struthiosaurus* and *Hungarosaurus* to Europe, as indicated by the bar

to be the result of speciation within Europe (figure 7.8). In addition to this European source area, there are two migrations: one to South America, involving *Rocasaurus* and *Saltasaurus*, and the other, for *Malawisaurus* and *Rapetosaurus*, to Africa.

Other than the dinosaurs, biogeographic affinities can be worked out for several other Transylvanian taxa, including *Allodaposuchus*, *Kallokibotion*, and kogaionids. Based on its phylogenetic position,[35] *Allodaposuchus* appears to have descended from a Euramerican common ancestor shared with *Hylaeochampsa*, another basal eusuchian from the Early Cretaceous of England. This relationship, and their stratigraphic occurrences, indicate a long ghost lineage (55–60 million years) for *Allodaposuchus*. Furthermore, its distribution appears to be the result of insular dispersal combined with the Euramerican breakup. What is more significant is the possibility that, with *Allodaposuchus*, *Hylaeochampsa*, and *Iharkutosuchus* (a eusuchian from the Late Cretaceous of Hungary)[36] all having a European distribution, crown-group crocodilians

Transylvania, the Land of Contingency   179

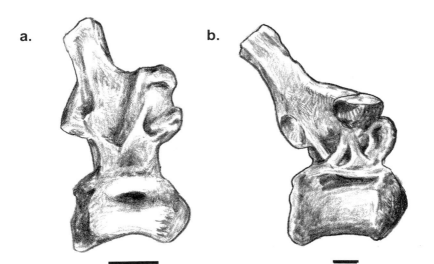

Figure 7.7. (a) A dorsal vertebra of *Lirainosaurus astibiae* from the Late Cretaceous of Spain, in right lateral view; and (b) a dorsal vertebra of *Ampelosaurus atacis* from the Late Cretaceous of France, in right lateral view. Scale = 5 cm. ([a] after Sanz et al. 1999; [b] after Le Loeuff 1995)

may also have dispersed from Europe to North America sometime during the Early Cretaceous.[37] The phylogenetic position of the Transylvanian turtle *Kallokibotion* is somewhat controversial. In some studies, it has been placed as a basal cryptodiran most closely related to *Tretosternon*.[38] The lineage leading to *Kallokibotion* therefore must have had its origin by the earliest Cretaceous (the oldest occurrence of the *Tretosternon* clade). Because it represents the Late Cretaceous terminus of an old phylogenetic lineage, *Kallokibotion*, with a 70 million year ghost lineage, constitutes an endemic relict in Europe, due to the breakup of Euramerica. Other analyses have situated *Kallokibotion* outside crown-group Testudines.[39] This more basal position does not alter the hypothesis of European endemicity. However, shifting its position more basally within Testudinata drastically increases the ghost lineage duration leading to *Kallokibotion* (by about 20–25 million years, from the Middle Jurassic to the Maastrichtian). Finally, kogaionid multituberculates are known solely from the Late Cretaceous of Transylvania and the Paleocene of France, Spain, Belgium, and Romania.[40] Optimization of the distribution patterns reveals that this clade of mammals had a European

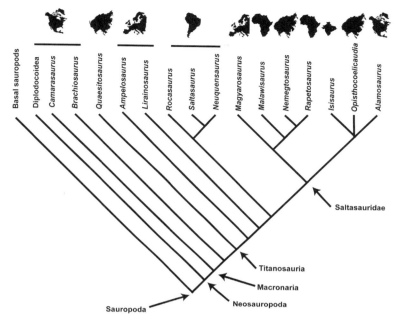

Figure 7.8. A simplified cladogram for basal ornithopods that also includes geographic information, indicating the dispersal of *Magyarosaurus*, *Lirainosaurus*, and *Ampelosaurus* to Europe

origin, stemming from a broader Euramerican distribution, sometime during the Early Cretaceous.[41]

Other Late Cretaceous localities in Europe lend themselves to similar interpretations for the other Transylvanian dinosaurs. Do these Spanish, French, and Hungarian faunas support, or perhaps alter, our earlier biogeographic interpretations for other Transylvanian dinosaurs, such as in the case of *Telmatosaurus*? We can't say. We know far less about the biogeographic dynamics of the Transylvanian theropods, because we know next to nothing about the details of their respective phylogenetic relationships.[42] How they will turn out, we don't yet know; only future discoveries and analyses will tell.

Clearly, we have a long way to go to grasp the full biogeographic dynamics of the dinosaurs from Transylvania and elsewhere in Europe. Yet, from what can be discerned so far, there is growing evidence that the biogeographic history of the Transylvanian taxa is more complex than meets the eye, and certainly much more so than Nopcsa had imagined. Here's what our present data tell us. Two lineages endemic to Europe—one leading to *Struthiosaurus* and the other to Rhabdodontidae—diversified into small clades there, one recognized at the species level (within *Struthiosaurus*) and the other at the generic level (within Rhabdodontidae). They all became extinct in Europe slightly before or at the end of the Cretaceous, without giving rise to new taxa. They were the headstones marking their respective clades, never to tread elsewhere in the world.

For the titanosaur clade of *Magyarosaurus*, *Rapetosaurus*, *Malawisaurus*, and *Nemegtosaurus*, only the first-named remained in Europe, whereas the second two migrated to Africa, and the last-named to Asia. Thus Europe acted as a partial venue of diversification, with *Magyarosaurus*, the last titanosaur from this part of the world, becoming extinct just before the end of the Cretaceous.

If our resolution of the cladogram is correct, then *Telmatosaurus* stands apart from these Transylvanian dinosaurian dead ends. First, this individuality was the product of a single ancestral migration to Europe by a North American or Asian iguanodontian. Second, this ancient invader likely gave rise to a modest hadrosaurid diversification, in order to account for their ultimate Transylvanian, Italian, and Franco-Iberian distribution in Europe. However, according to their phylogeny, the legacy of this European nexus of primitive hadrosaurids also spilled back into North America and into Asia. The great radiation of euhadrosaurians came with their migration *out* of Europe, with important consequences for their anatomy, development, and evolutionary dynamics, which we will examine below.

In summary, the common ancestors of the immediate clades of *Zalmoxes* (with *Rhabdodon*) and *Struthiosaurus* (with *Hungarosaurus*?) migrated to Europe from North America, *Telmatosaurus* dispersed outward from Asia or North America, and *Magyarosaurus* differentiated itself from other European titanosaurs, all probably in the Early Cretaceous or Late Jurassic (chapter 4). If so, then each colonization took place as continental configurations were becoming ever more complex. Beginning in the Early Jurassic, northwestern Africa and eastern North America had drifted apart to form the beginnings of the North Atlantic Ocean, but this separation occurred at what today would be the eastern seaboard of the United States to the west and Morocco and the adjacent coastline to the east. The northernmost part of this proto-Atlantic Ocean was only just beginning to open, whereas the southern part, between Africa and South America, would not open until several million years later. Thus intermittent land connections must have existed before and into the Early Cretaceous (about 100 million years ago)—between what is now Labrador in northeastern Canada, Greenland in the middle, and Europe to the east—through a dense array of large islands. Thereafter, the terrestrial habitats associated with these landmasses came and went as the tectonics of the region changed and the sea level fluctuated. At the same time, possible dinosaurian dispersal from Asia was also limited, by sporadic land bridges across the Polish Trough (a large but intermittent seaway running southeast from the present Baltic Sea and across Ukraine), and by the West Siberian Sea and Turgai Straits (extending north of the present-day Caspian Sea to the paleo-Arctic region),[43] both of which existed from the Middle Jurassic to the Oligocene.[44] To the south was a string of islands of various sizes that stretched from what is now central Asia to Anatolia.

The time periods appear to fit the hypothesis, so our conjecture that the initial introduction of the ancestors of *Telmatosaurus*, *Zalmoxes*, and many of the rest of the Transylvanian dinosaurs from North America or Asia occurred sometime in the Late Jurassic through the Early Cretaceous seems to be warranted, at least based on current evidence. But why was it these taxa and not a different set of migrants? Where are diplodocid sauropods, allosaurid and therizinosauroid theropods, and stegosaurs, all of which were present in North America and Asia over this stretch of time? Why the gamut of dead ends and success stories for the Transylvanian dinosaurs themselves? These questions are easy to ponder, but hard to answer. Is there any reason why the ancestors of the Transylvanian dinosaurs should have been more likely to leave North America or Asia to end up in Europe than any of those left behind? Why should the ancestry of *Telmatosaurus* also have spawned the great euhadrosaurian radiation elsewhere in the world, but that of *Struthiosaurus*, *Zalmoxes*, and *Magyarosaurus* remained so barren and restricted? We're hard pressed to explain these patterns (either individually or collectively) as the results of some common property that enhanced their ability to migrate or colonize, a characteristic that the "left-behinds" wouldn't have possessed. The travelers certainly don't appear to have been more honed by natural selection for long-distance, multienvironmental journeys—which would have increased the probability of colonization—than the other dinosaurs of those times. Instead, who stayed and who went is unpredictable, reflecting historical contingency. Only by chance, and not by choice, did the progenitors of *Telmatosaurus*, *Zalmoxes*, and the rest manage to enter the mosaic of microcontinental movement and of submerging and emerging landmasses that is now Europe. Being in the right place at the right time was no more than luck.

A RETURN OF THE DWARFS

In this geographic and ecological context, we now return to our miniaturized Transylvanian dinosaurs. For these heterochronic dwarfs, small things (literally and figuratively) can lead to big things in unpredictable ways. A miniaturized dinosaur, free of the constraints imposed by a large body size, has more versatility to adapt to a changing environment by means of evolutionary innovations.[45] Using several examples from Transylvania, we will examine how, when, and why particular features of these dinosaurs were either predictable or the random consequences of their changing statures, as well as consider how they may fit into the evolutionary history of their respective clades.

As we saw in chapter 5, *Telmatosaurus*, *Magyarosaurus*, and possibly *Struthiosaurus* were identified as being dwarfed, downsized from the primitive condition of their respective clades. In addition, *Zalmoxes robustus* (but not *Z. shqiperorum*) may have been dwarfed. However, the Transylvanian theropods are roughly the same size as elsewhere; the same applies to the region's crocodilians, turtles, and mammals. We also noted the possibility of dwarfing in some of the Transylvanian pterosaurs, but certainly not in *Hatzegopteryx*, one of the largest among the large azhdarchid pterosaurs.

By integrating this cast of characters with their evolutionary relationships, and with their geographic and stratigraphic distributions, we can now begin to explore the biogeographic dynamics by which they achieved their dwarfed status. Sometime in the Early Cretaceous (or earlier), the ancestor of each of the Transylvanian lineages crossed the northern realm between North America or Asia and the mosaic of large and small landmasses that formed what is now Europe. We suspect that these migrants were no smaller than their closest stay-at-home relatives—since fasting endurance is proportional to body size,[46] only large individuals would have been able to travel long distances without risking starvation, famine-induced disease, or susceptibility to predators.

Some of these migrants colonized the isolated, rapidly changing patches of terrestrial habitats in Europe. This region—isolated as it was from the grand coastal plains from which they came—was a place of serendipity, the product of the chance arrivals of various plants and animals. What were once more-or-less stable, evolutionarily well-honed ecosystems in North America and Asia appear to have been stitched together more by chance than by design as they developed in Europe. One hallmark of the kind of isolated habitats we envision here is the taxonomic and trophic imbalance of the European faunas.

Things now are starting to get really interesting, once each migrant had become established in these new habitats. Whereas those who traveled to Europe seem to be a random sample from their ancestral area (although they needed to be large enough in body size initially), selection appears to have favored downsizing. Very quickly, being large was hardly as advantageous as it had been before and during the journey to Europe. Many attempts have been made to account for dwarfing in isolated organisms. For example, large body size requires lots of resources—which were relatively plentiful in the now-forfeited coastal plains of Asia and North America, but perhaps critically rare in the new homeland. From this metabolic perspective, it would no longer pay to be big. Instead, it is reasonable to expect that the size reduction in these animals, living in restricted European habitats, was due to selection for small body size, based on resource limitations.[47]

From an ecosystem perspective, downsizing could also have been related to a release from predator pressure that comes from the chance jumbling up of colonizing predators and prey (chapter 5). The randomness of which predators arrived and evolved in Transylvania and elsewhere in Europe and which ones did not would have unpredictably altered the predator-prey relationships in these new regions. Selection for large body size as an antipredatory device, which may have existed in Asia and North America, would have been out of whack in Europe. It has been argued that with a loss of or decline in body-size selection, the distribution of body sizes in prey may evolve into an equilibrium that is not mediated by predator pressure—large prey get smaller and small prey become larger when they're not using size to escape from their usual predators.[48]

Even though both of these ecological explanations of dwarfing are good enough by themselves, we argued in chapter 5 for the Roth model of downsizing: dwarfing was the product of selection based on life histories in an isolated setting.[49] In this model, only a few individuals ever disperse to any isolated region. The resulting population bottleneck entails an immediate loss of genetic variation in the organisms colonizing the newly invaded environments. This small population is usually doomed to extinction through the consequences of genetic drift. However, in order to reduce the probability of extinction, a premium is placed on the invaders to increase their population size as rapidly as possible. Options on how this can be achieved are limited, but selection for early (precocial) sexual maturation—one aspect of what is called r selection— may be the colonizers' best chance for success.[50] By moving up the timing of its sexual maturity, an organism will be in a better position to increase its presence in the environment more quickly than its slower-developing ancestors and competitors. While this is applicable for whatever the quality of the colonized environment, it is even more effective in increasing survival rates in an ecologically unstable environment.[51] In addition, precocial species are also better at exploiting patchy resources, and they disperse across their new habitats earlier than their more slowly developing competitors do.

Under this life-history approach, r selection, in the context of colonizing isolated regions, acts primarily on accelerating sexual maturity—a good colonizer is one that can quickly return to a healthy population size after the migrational bottleneck. Interpreted this way, dwarfing is not the direct consequence of selection, but instead is a tagalong, a predictable consequence of selection for those life-history parameters that relate to colonizing abilities.[52]

For our Transylvanian dinosaurs, early sexual maturity certainly accounts for the evolution of their small body sizes, especially in the context of their arrival in Europe. By means of their early maturation, they would have been better off than their larger ancestors during this time of population bottlenecking, simply by being able to recover better and increase more quickly in population size. These factors probably created a feedback loop with the other features that favor downsizing, such as limited space and resources and new predator-prey relationships. Taken together, the resulting morphology is an animal smaller than its ancestors.

How long it may have taken for this dwarfing to occur is hard to reconstruct from the fossil record. In other vertebrates, for instance in the dwarfed hippos of Crete, it is thought that dwarfing was achieved within a millennium, or even shorter periods of time.[53] We don't know if this time frame also applied to the Transylvanian taxa, due to our lack of information (concealed by their long ghost lineages). Nevertheless, we can reasonably conjecture that the isolation of these dwarfed populations, over whatever interval of time, not only reduced their genetic variability (by reducing the influx of expected variation, due to their migration), but also probably accelerated the divergence of the dwarfing populations from their mainland stocks.

AFTER THE DWARFS

Living as dwarfs in isolation may not have been all it's cracked up to be, at least in terms of evolutionary legacy. For all but one of the Transylvanian dinosaurs, we are looking at the final pulse of their immediate clades. The likes of *Zalmoxes* and *Struthiosaurus*, in their isolated outpost within Europe, died out at the end of the Cretaceous without giving rise to other taxa. *Magyarosaurus* was part of only a small evolutionary radiation. Only in *Telmatosaurus* can we identify a link between dwarfed isolates and their subsequent migration and high levels of diversification. If we want to examine the great radiation of euhadrosaurian species in North America and Asia from non-hadrosaurid iguanodontians, we've got to go through *Telmatosaurus* and its European allies. In doing so, we hope to explain how this European occurrence of basal hadrosaurids is an important, but historically contingent factor in the rise of Euhadrosauria.

The construction of the hadrosaurid dentition is distinct from that of more primitive iguanodontians such as *Tenontosaurus*, *Dryosaurus*, *Camptosaurus*, and *Iguanodon*, so much so that it has been considered a hallmark for Hadrosauridae (chapters 2 and 5). Instead of having a dentition composed of a single large functional tooth and one replacement tooth per tooth position, hadrosaurids had hundreds of small functional and replacement teeth in each jaw, interlocked to form a complex dental battery that is ornate by anyone's criterion (figure 2.21). Each tooth was composed of various kinds of dentine tissue and contained much less of the harder and more resistant enamel; yet, when worn down, this dentition provided a long, roughened surface to grind up tough plants, a dentition more effective at breaking down fibrous leaves than the teeth present in their ancestors. Teeth were continually replaced, such that this chewing surface was always present throughout the animal's life. When the teeth in both the upper and lower jaws were worn, the opposing occlusal surfaces formed a complex arrangement of enamel and dentine, wonderfully designed to grate and rasp their way through the toughest plant material in ways not available to hadrosaurid precursors.

This dental battery certainly was a remarkable trophic invention, one that improved hadrosaurids' chewing efficiency over the more primitive condition of mastication occurring among ornithopods. Surely it qualifies as a true adaptation—that is, an organismal feature whose origin and maintenance was the product of natural selection.[54] Viewed in terms of the Cretaceous rise and diversification of flowering plants, a dental battery makes sense from such a Darwinian perspective: new kinds of plants provide new challenges (i.e., ecological problems and opportunities) for contemporary herbivores, which then select for those variants among iguanodontians that have an incipient dental battery (i.e., biotic solutions to these problems and opportunities). Using these arguments, it is always tempting to take an adaptationist perspective, rendering the rise of the hadrosaurids in terms of their coadaptation or coevolution with angiosperms.[55]

However attractive this inference is, we're not particularly persuaded by this adaptationist explanation, whereby the evolution of the hadrosaurid dental battery was directly mediated by natural selection. We think it appropriate to look first at alternative explanations, using all the tools, perspectives, and information we now have at hand. In particular, that means an examination of dwarfing and migration to the isolated regions of Europe.

Dwarfing within the lineage of iguanodontians that gave us *Telmatosaurus* and the other basal hadrosaurids of Europe brought about the possibility of adding a few more teeth in the tooth row and another replacement tooth per position—the beginnings of an incipient dental battery. With this more complex dentition, made up of small teeth placed in small jaws, it is possible that these dwarfed hadrosaurids altered what they fed upon. As has been hypothesized for dwarfed, island-dwelling elephants, the Transylvanian dwarfs are likely to have experienced a decrease in their total metabolic requirements, coupled with a reliance on higher-quality food, assumptions based on their reduced body sizes and decreased population size.[56] Whether the European ancestors of *Telmatosaurus* subsisted on a diet higher in nutrient-rich seeds and fruits than their larger North American and Asian relatives is unknown, but this may be anticipated, since dietary selectivity in many living animals is known to increase with smaller body size.[57]

Whatever the diet of these miniaturized hadrosaurids, their downsized stature—a consequence of early maturation—is likely to have promoted their success in the more restricted terrestrial habitats of Europe. Nonetheless, according to their phylogenetic history, the descendants of these animals migrated from Europe (most likely first to North America, but an Asian migration is also a contender here), going from a suite of environments that promoted small size back to a region of great coastal-plain openness. As far as we can tell, this out-of-Europe migration happened once, to produce what we now call euhadrosaurians (chapter 2).[58] Once they were "back home," their small size could not have been as advantageous as it had been in their restricted European habitats. These dwarf descendants would have again faced a great array of small and large predators (dromaeosaurids, tyrannosaurids) already present in North America and Asia. It's possible that through the kinds of predator pressure their more distant ancestors experienced, they regrew to their previous larger sizes as a way to defend against predation. Whatever the case, these euhadrosaurians once more reached 10 m or more in length.

As far as we can tell, this size-increase phase equally affected just about all of their body proportions (as exemplified by *Telmatosaurus*). All, that is, but their dentition (chapter 4). Rather than scaling up to a bigger size (like the rest of the body), the teeth remained small, not much larger than those of hatchlings and youngsters. In other words, euhadrosaurians retained their dwarfed, juvenilelike dentitions into adulthood. In this way, hadrosaurids appear to have decoupled the increase in their overall body size from their dentition. Mediated by their European dwarfing phase, the teeth of these North American and Asian hadrosaurids, as they rebounded to a larger body size, paedomorphically remained small. Simply add more of these miniaturized teeth (i.e., baby teeth) along the length of their jaws and as replacement teeth, and voilà, the hallmark dental battery for which Euhadrosauria is famous had been created.

Having a dental battery must have counted for something among all hadrosaurids. It is retained in all of the nearly 50 species of euhadrosaurians presently known worldwide, never deviating in its fundamental construction—small teeth, with three or more teeth per tooth position. Once formed, this dental battery became one of the most stable of all euhadrosaurian anatomical systems. From an engineering perspective, this dentition appears to have been a great invention, permitting the complex mastication of plant material. Once this structural complexity, and therefore better chewing efficiency, was achieved, natural selection would have acted to *maintain* the hadrosaurid dental battery. Only in this sense can the stability of the hadrosaurid dental battery be considered an adaptation, and we certainly wouldn't be surprised if this was the case.[59] However, the *origin* of dental batteries cannot be judged from the same perspective. This evolutionary innovation arose not as the primary object of selection, but instead as a serendipitous creation. Hadrosaurid dental batteries came about from the transformation of the dentition of the numerous, more basal iguanodontians to that of *Telmatosaurus* and other hadrosaurids, not for their properties related to chewing, but through the sequellae of paedomorphic dwarfing.

This notion of contingency in the evolution of hadrosaurid teeth may have its mirror in what we know about hadrosaurid reproductive biology. From work by Jack Horner of the Museum of the Rockies in Montana, we know that hadrosaurids such as *Maiasaura* and *Hypacrosaurus*— much larger forms than *Telmatosaurus*—laid large clutches of relatively small spherical eggs (between 16 and 22 eggs in a clutch, with the egg volume ranging from 900 to 4,100 cm^3) and cared for the small (approximately 1% of adult body mass), immature hatchlings for some length of time until the juveniles left the nest (figure 7.9).[60] Parental care of immature offspring is usually selected for in populations at or near the carrying capacity of the environment, what is known as the K life-history strategy. The logical opposite of r strategy that we previously discussed, K strategy is to be expected in animals of large adult body size inhabiting stable environments of the kind envisioned in western North America

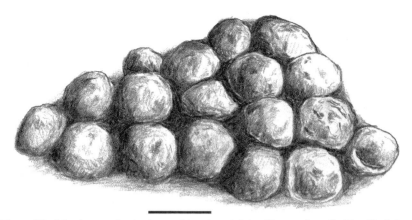

Figure 7.9. A lambeosaurine hadrosaurid nest containing 18 eggs, from the Two Medicine Formation (Upper Cretaceous) of western Montana. Scale = 10 cm. (After Horner and Dobb 1997)

and Asia during the Late Cretaceous.[61] K selection is also associated with small clutch size and parental care. In this way, the parents are able to invest a great deal of their energy promoting the survival of a few offspring. Humans, for example, are regarded as K strategists. However, a large clutch size is more consistent with an r strategy, in which the parents invest in abundant offspring, but do not care for them. Among vertebrates, sea turtles are an excellent example of this strategy. They lay plenty of eggs in the sand, but care for them not at all. After climbing out of their eggs, the hatchlings must fend for themselves.

Hadrosaurids exhibit both r and K strategies in their reproductive biology. Is this a paradox, requiring an explanation in which natural selection is driving life-history traits in polar-opposite directions? Not necessarily, if we interpret these reproductive features along the lines of our schema of a European dispersal and dwarfing at the base of the hadrosaurid family tree. Recall that early maturation and its consequent reduction in body size make for good colonizers. In addition, selection on individuals in populations well below the carrying capacity of their environments favors the early and rapid production of large numbers of young. This aspect of r strategies makes good sense in such isolated terrestrial realms as Transylvania in the Cretaceous. Large clutch size is likely to have evolved in the context of colonizing the unpredictable, often severely fluctuating Neotethyan terrestrial habitats. The large clutch size of small hatchlings that we see in euhadrosaurians would then be primitively inherited from the isolated origin of the group, an evolutionary holdover from more stressful times. The K strategies (stable habitat, large body size, parental care) would then have evolved after hadrosaurids left Europe for North America and Asia. Now known as euhadrosaurians, these creatures evidently maintained their earlier r-selected reproductive strategies in much the same way as they ended up with their miniaturized dentitions.

In sum, unpredictability abounds in the details of what we've presented here. If we are correct about how the downsizing of the Transylvanian dinosaurs was accomplished, then it would have been impossible to predict from their North American or Asian ancestors which taxa would make the journey and, once there, which ones would have been successful in the restricted environments of Europe. These colonizers then became dwarfed, not so much from a direct selection for body size, but instead as the consequence of selection for early sexual maturity.

Populations eking out a living in isolation, even with their odd genetic sampling and high levels of speciation, usually end up as evolutionary dead ends. This may have been true for the Transylvanian dinosaurs—Europe was their last stop. However, *Telmatosaurus* and its closest European relatives were the progenitors of the euhadrosaurians on the larger adjacent continental landmasses. Their reintroduction into Asia and North America (the postdwarf stage) has its own contingency overprint, with the same iffy-ness of dispersal and success or failure in these descendant regions. In this group, the effects of isolation on life histories indirectly produced evolutionary novelties, such as the hadrosaurid dental battery. These features were then bequeathed to descendant species that evolved into a diverse array of larger animals. In other words, the period of r selection experienced by the basal taxa of Hadrosauridae, which was related to their early sexual maturity in an uncrowded insular habitat, supplied a pool of dwarfed morphology that could be used as the raw material for the subsequent radiation of these herbivores. If true, then here we encounter one of the few examples of isolated populations successfully recolonizing, radiating, and dominating the biota in a broad continental setting.[62]

Our ability to identify serendipitous historical conjunctions—changes in features, resulting from a different interactive context—provides us with the means to suggest alternatives to explanations that involve immediate selection for biological features. Such momentary shifts in context can have huge consequences, some fatal, others with no effect at all, and still others that are spectacularly successful. Whatever the case, these shifts and their consequences cannot be predetermined, or even be strictly predictable; they can only be understood in a historical sense.[63]

CHAPTER 8

Alice and the End

Ancient Transylvania now makes sense as a land of contingency, a region where any number of unpredictable events provided the raw material for what would become the history of members of its dinosaurian fauna. Emerging from a synthesis of cladistics, paleogeography, heterochrony, and life-history strategies, the significance of the Transylvanian dinosaurs comes from what their evolution tells us about historical contingency. It's even more exciting when studies like this transcend their intrinsic appeal to provide insights into larger issues in evolutionary biology. We think maybe this has happened here. To get there, however, we should probably first revisit one of the finest adventures to come from children's literature. Enter Charles L. Dodgson, a British mathematician, deacon, and photographer, but better known throughout the world as Lewis Carroll, the author of *Alice in Wonderland* (1865) and *Through the Looking Glass* (1872).

Through the Looking Glass begins with Alice dreaming of being in a house where everything was in reverse, left to right, as if she had passed through a mirror. Outside was a beautiful garden, but she wanted to see it better from a nearby hill. However, as she followed what appeared to be a very straight path to the hill, she found that it led her back to the same house. When she tried to speed up, she instead immediately returned and crashed into the house. Eventually, Alice found herself in a patch of very vocal and opinionated flowers, telling her that someone (the Red Queen) often passed through this garden. When the Red Queen finally came into view, Alice tried to catch up with her, but to no avail; the Red Queen quickly disappeared. One of the roses in the garden advised Alice to walk in the opposite direction and, by this reversed action, she immediately came face-to-face with the Red Queen.

As the Red Queen led Alice directly to the top of the hill, she explained to her that in this world, hills can become valleys, valleys can become hills, straight can become curved, and progress can be made only by going in the opposite direction. At the top of the hill, the Red Queen began to run, faster and faster. Alice, following after the Red Queen, was further perplexed to find that neither one of them seemed to be moving. When they stopped running, they were still in exactly the same spot. Alice remarked on this, to which the Red Queen responded: "Now, *here*, you see, it takes all the running *you* can do to keep in the same place."

We haven't the slightest idea whether Lewis Carroll gave much thought to the evolutionary implications of Alice's adventures with the Red Queen, although it might have been possible—Carroll lived no more than 45 km from Darwin's Down House home in the neighboring county of Kent, and the two published their works just 13 years apart (*On the Origin of Species* was first released in 1859). Nevertheless, it took a century, and travel nearly 7,000 km to the west, for Carroll's Red Queen to provide the moniker for a major shift in our understanding of evolutionary dynamics.

In 1973, Leigh Van Valen (figure 8.1), an evolutionary biologist and vertebrate paleontologist at the University of Chicago, published a paper that explored whether long-lived lineages were more resistant to extinction than short-lived groups. According to Darwin, natural selection weeds out the poorer- and promotes the better-designed organisms, thus improving the fit with their environment. If so, Van Valen then reasoned that a species should have less probability of becoming extinct in the future if it has already been around for a while. In other words, long-lived species are presumably the ones that have been finely tuned by natural selection. To test this proposition, he examined the rates of extinction for a large number of evolutionary lineages, compiling what amounts to life tables for a variety of organisms, from clams to mammals.[1]

To Van Valen's surprise, his data showed that the probability of extinction remains constant, independent of whether the species has been in existence for a long or a short time. In order to explain this seeming

Figure 8.1. Leigh Van Valen (b. 1935), portrayed as Alice being pulled through an ever-moving landscape by the Red Queen. (After the original by John Tenniel in Lewis Carroll's 1871 *Through the Looking Glass*)

paradox, he argued that a species' state of adaptation must be under continual assault by an environment that itself is undergoing constant change. Van Valen likened this lack of staying power on the part of long-lived species to Carroll's Red Queen. In the same way that the Red Queen must perpetually run in order to keep up with her literal place in the world, so, too, must organisms constantly evolve in order to simply maintain their state of adaptation. The Red Queen hypothesis (also referred to just as the Red Queen) should be especially applicable to cases with tightly coevolving species. In these instances, where the evolution of one species is directly related to its interactions with another (the biotic environment), the effect of these coevolutionary relationships is that as one species adapts to its environment and thus gains a competitive advantage over other species, the latter necessarily become less well adapted. The only way that a species in this the latter group can survive is by responding with its own improved design. From there on in, the emphasis on improvement becomes increasingly reciprocal; it can even go so far as to produce an evolutionary arms race. Caught in a zero-sum game of competition, the best a species can do is to keep producing evolutionary innovations (running) to try to tip the scale in its favor, just to maintain its existence (to stay in the same place), the whole time that its coevolutionary partner is trying to do the same. To win is to keep playing this game. Otherwise, that species is an evolutionary loser, having failed to keep up and therefore doomed to extinction.

The Red Queen has been applied most successfully to the coevolution of hosts and parasites, predator-prey systems, and even to the persistence of sexual reproduction.[2] The arms races between the red-backed shrike (*Lanius collurio*) and common cuckoo (*Cuculus canorus*), between the predator gastropod known as the whelk (*Sinistrofulgur*) and its dangerous bivalve prey (*Mercenaria*), and between bacteria and bacteriophages, all ensure that each individual species' state of adaptation is a transient thing[3]—in a constantly changing environment, natural selection operates to enable organisms simply to maintain their state of adaptation, rather than improve it. The logical underpinning of the Red Queen hypothesis is a strict reliance on natural selection, which predicts that the best design/adaptation for one organism will be thwarted by tightly interlocked, interspecific competition with other organisms.[4] Thus, because participants are a part of each other's biotic environment, the feedback loop of selection between them produces the pattern of extinction documented by Van Valen—the probability of extinction is no different for old or young clades.

Our interest in identifying how chance events can influence evolutionary history grew as we developed our interpretation of the Transylvanian dinosaurs, where we focused on the creative possibilities inherent in random events. We also began thinking about the original explanation of the Red Queen and its reliance on interspecific selection as the directing force in evolution, with predictable consequences. We wondered if, perhaps, the Red Queen curves may have an entirely different message, one going beyond elucidating the relationship between adaptation and extinction among coevolving lineages.

Let's begin by looking at the sources of perturbations that keep organisms at suboptimal adaptation. For biotic systems, the Red Queen envisions a constant evolutionary arms race between species, with one, and then the other, temporarily holding sway in their coevolutionary competition. Perturbations come from the evolution of new features in one half of the arms-race pair, momentarily improving their fitness, which is subsequently overturned by their co-competing partner. This is all happening under the umbrella of a host of abiotic transformations—such as climate change, tectonic events, and meteoritic impacts (what

Alice and the End

Tony Barnosky of the University of California at Berkeley has called the Court Jester[5])—that are also constantly acting to dampen the process of adaptation.

From an organism's perspective, it is probably irrelevant whether the environmental alterations are biotic or physical; they have to be dealt with one way or the other in order to survive. If these two kinds of impacts are indistinguishable to a species, then perhaps what matters most in determining the Red Queen curves is that, wherever the disturbances come from, they should be random with respect to the organism's current state of adaptation.[6]

Looked at from this different angle, the Red Queen can be seen as more than a theory about the relationships among coevolutionary selection, adaptation, and extinction. It now becomes a testimony to the ubiquity of random events of all kinds, as well as to their unpredictable consequences—both positive and negative—in evolutionary history. The bulk of this book has been dedicated to identifying the contingent opportunities derived from the unexpected European migration of the ancestors of several dinosaurs. Such unpredictability, in turn, may also have put an end to them individually—or, more strikingly, *en masse*—during the great Cretaceous–Tertiary extinction 65 million years ago.

Until recently, our explanations of why life is the way it is has been left in the hands of natural selection; because of their ahistorical and lawlike nature, adaptive hypotheses always trump alternatives. These explanations are most often couched as virtual pathways by which features evolve in a deterministic fashion under the constraints provided by physics, self-organization, and other constructional/developmental principles. These inferences may satisfy, but determining the potential track of natural selection is not necessarily equivalent to reading history. What do we do if an evolutionary pattern based on selection belies the homologies used to construct the history of these organisms? Clearly, we need a tool to separately assess the roles of contingency and natural selection during the unfolding of history, and, in this book, we have used phylogenetic analyses to identify this historical context.

Identifying unpredictability in history, however, is a double-edged sword. By decoupling history from determinism, we alter our ability to rely solely on the laws of nature to test biological questions. We can't do without this background of natural laws. Some boundaries of these laws are stringent, with very narrow channels that restrict the range of possible events, whereas others are vaguer and the pathway for potential events is broader. On the other hand, historical contingency—ultimately affecting this physical background—provides a role for both creativity and destruction in producing the extraordinary richness of nature.

To reconstruct and interpret the evolution of life on Earth, however, we need to set aside our reliance on selection for or against particular features as the sole explanation for their existence, and instead examine the role of contingent vagaries of history on our understanding of how evolution has worked. If, as Gould has suggested, evolutionary history is ruled by unpredictability, then nature may not be a zero-sum game eked out on a coevolutionary stage, as has often been proposed. Instead, it may be that odd, minor, unpredictable events of a physical kind that are faced by organisms are as effective as biological arms races in producing environmental decay with respect to the adaptive state of existing organisms. Such historical contingency—and not just coevolutionary selection—leads to the Red Queen, catapulting organisms on equally unpredictable pathways. In this way, the unpredictability of natural history can opportunistically soften or skirt the laws of nature, and thereby provide greater possibilities for further contingency.

C'EST FINI

Throughout our work on this project, we constantly asked ourselves why anyone should be interested in this stuff. After all, it's a long leap from digging a bone or tooth out of the golden-brown rocks outside the village of Sânpetru on a hot day to historical unpredictability and Alice and the Red Queen, with many uncertain twists and turns from beginning to end. Perhaps because of this long (and, shall we say, contingent) jump, we became fixed on this self-reflexive question of ours, which, in the end, we thought was related to what might be called science as a game—not a game full of strict rules, but one of enjoyment.

We questioned why, in the first place, we indulged our penchant for these dinosaurs and the issues to which they have led us. Unabashedly, our initial motivation was that it would be fun and exciting. And why not? We obviously thought (and continue to think) that the prospects of our endeavor, ranging from fieldwork to aspects of evolutionary theory, looked fascinating. So we got on board to see where it would take us.

Beyond amusement, or perhaps as an extension of it, what we got out of playing this game fed back on itself, making the game more fascinating still. Our love of the subject and what it reveals about the ways of the world were the lynchpins that kept us participating in the project. This also led us in often novel and bizarre directions, forcing us into situations where we had to confront our ideas about a variety of scientific and philosophic matters. Some of these directions proved to be uninteresting, redundant, or barren, and they were subsequently dropped. Ultimately, though, whether the body of this work and its tangents are proven right or wrong, or declared insightful or misleading, are all irrelevant to our original decision to play the game we embarked on. Our goal was to try to make an honest effort at constructing a story about the dinosaurs of Transylvania, on as rich, to the point, and true to the details as we know them. Our intent was to expand our pursuit widely and delve deeply into bones, teeth, and rocks, applying parsimony to character evolution, using a host of ecosystem information, and following the choreography of continents through geologic time. As a result, you might think that we've arrived at the truth about the Transylvanian dinosaurs and their place in evolutionary history, but we cannot pretend that this is the case, nor that it was our goal or our motivation. Have we indeed found out what, exactly, happened to produce the Transylvanian dinosaurs? We doubt it. These dinosaurs sparked our efforts, which we took seriously as we progressed, but what we gleaned from all of this should not necessarily be taken as the truth. In other words, this game of science isn't won by revealing what "really happened"; its purpose is as much as in providing the enjoyable possibility of allowing one to keep playing the game.

So why should anyone be interested in our prehistoric journey? From our own perspective, it ought to be because of its intrinsic fascination, leading to ever-increasing ripples in how one views the world. This is why we think that research such as what we've discussed here—abundantly evident in so many of the studies being conducted today—has great value to our understanding of and excitement about prehistory. It relies on the challenges of hard work, insight, imagination, and skepticism but, ultimately, rests on the availability of fossils, those meager resources that are also the only physical vestige we have of former life on this planet. Here we've tried to make the most of the extremely rare material that nature has provided, both to indulge ourselves and maybe to infuse you with some of our own fascination.

In 1964, Marcel Duchamp—one of the twentieth century's most influential artists—spoke with his biographer, Calvin Tomkins, about his views of science:

> We have to accept those so-called laws of science because it makes life more convenient, but that doesn't mean anything so far as validity is concerned. Maybe it's all just an illusion. We are so fond of ourselves, we think we are little gods of the earth—I have my doubts, that's all. The word "law" is against my principles. Science is so evidently a closed circuit, but every fifty years or so a new "law" is discovered that changes everything. I just don't see why we should have such reverence for science, and so I had to give another sort of pseudo-explanation. I'm a pseudo all in all, that's my characteristic. I never could stand the seriousness of life, but when the serious is tinted with humor, it makes a nicer color.[7]

It is in Duchamp's light that we view our contribution on the dinosaurs of Transylvania. Counterposing the capricious nature of history against what we call the "laws of nature" ultimately reduces the overarching command of the long arm of these laws in determining history. Our efforts have left sundry questions unanswered: many because we haven't the ability to answer them, others because there's just not enough time in the day. Yet the rewards we've received from the science we've described here lie in the possibility of continuing to play. Our efforts to construct a good, bad, or indifferent story about the dinosaurs of Transylvania now stand on the brink of the next phase of the game. Like the 1918 *Dada Manifesto* of Tristan Tzara, Romanian-born founder of the Zurich Dada movement, we have "throw[n] up ideas so that they can be shot down."[8]

NOTES

CHAPTER ONE. *Bringing It All Back Home*

1. Boil 3 cups of water, add 2 tsp. of salt and 2 tbsp. of butter, add in 1½ cups of coarse- or medium-ground yellow cornmeal, reduce heat to medium, and stir for 10–15 minutes, until the cornmeal (polenta) thickens. Serve immediately with sour cream. Recipe from Klepper 1997.

2. A word should be said here about the various place names that are used in this book. Prior to the end of World War I, Transylvania was part of the Austro-Hungarian Empire, but it was transferred to Romania thereafter. Consequently, geographic and geologic names changed from Hungarian (or occasionally German) to Romanian. For example, Szacsal (Hungarian) is now replaced with Săcel (Romanian), Szentpéterfalva (Hungarian) with Sânpetru (Romanian), and Boldogfalva (Hungarian) by Sântămăria Orlea (Romanian). For additional information, see Nicolescu 1998.

3. Greene 1983; Suess 1916.

4. Nopcsa 1900. When originally described, this dinosaur was called *Limnosaurus*, but this name had already been used for an extinct crocodilian; in 1903, Nopcsa supplied the new genus name *Telmatosaurus*.

5. Buffetaut 1987; Colbert 1968; Weishampel et al. 1993.

6. Abel 1912, 1929; Weishampel and Jianu, unpublished manuscript; Weishampel and Reif 1984.

7. Edinger 1955.

8. Abel 1912, 1929; Nopcsa 1928a; Reif 1980.

9. Nopcsa 1900, 1903, 1905a, 1917a, 1917b, 1917c, 1918, 1922, 1923a, 1929a, 1929b, 1930.

10. Weishampel et al. 2004.

11. "You know the titanosaur material much better than I do. How would you like it if we worked on the remains from Transylvania together? You add the comparative-descriptive part and I add on top a paleobiological sauce, so to speak." Letter from F. Nopcsa to F. von Huene, 3 Oct. 1929. University of Tübingen, Friedrich von Huene archives (translation by Weishampel).

12. Nopcsa 1897, 1902a, 1905b, 1914a, 1914b, 1923b, 1926a, 1926b, 1926c, 1934.

13. For additional information on Nopcsa's activities in the Balkans, as well as on his family history, see Conrad von Hötzendorff 1921–1925; Corti 1936; Elsie 1999; Hamann 1986; Jianu and Weishampel 1998; Nopcsa 1905c, 1908, 1910, 1911, 1912, 1916, 1921, 1925a, 1925b, 1929c, 1932, 2001; Robel 1966; Weishampel and Jianu, unpublished manuscript.

CHAPTER TWO. *Dinosauria of Transylvania*

1. Grigorescu 1982; Groza 1982; Jianu 1992, 1994.
2. Codrea, Smith, et al. 2002; T. Smith et al. 2002.
3. Codrea et al. 2001, 2002, 2010.
4. The Late Cretaceous Transylvanian fauna provides the best record of dinosaurs in Romania, but there are two other sites in this country worth mentioning. The first, in eastern Romania (Cochirleni), yielded only a theropod tooth from the Early Cretaceous, referred to *Megalosaurus* cf. *superbus*. However, the second site, also from the Early Cretaceous in northwestern Romania (Cornet), has provided a wealth of dinosaur and other tetrapod material from a fissure fill in a bauxite mine. These specimens are nearly entirely disarticulated, dissociated, and abraded, rendering taxonomic referral difficult. Thus far, the recovered taxa include theropod, ankylosaur, and ornithopod dinosaurs, as well as pterosaurs; reported avian material is regarded as controversial.

For the Cochirleni "fauna," see Simionescu 1912. For the Cornet fauna, see Benton et al. 1997; Bock and Bühler 1996; Jurcsák 1982; Jurcsák and Kessler 1986, 1987, 1991; Jurcsák and Popa 1978, 1979, 1984; Kessler 1984; Marinescu 1989; Patrulius et al. 1982; Tallódi Posmașanu and Popa 1997.

5. Benson et al. 2008.
6. Turner 1997.
7. Hennig 1950.
8. Sober 1991.
9. Rüpke 1994.
10. Rüpke 1994.
11. Owen 1842.
12. Seeley 1887b.
13. Fastovsky and Weishampel 2009; Holtz and Osmólska 2004; Sereno 1999.
14. Fastovsky and Weishampel 2009; Sereno 1999; Weishampel 2004.
15. Le Loeuff 1997.
16. Nopcsa 1902b. See also Nopcsa 1904, 1905a, 1915.

17. Wilson and Upchurch 2003.
18. Huene 1932.
19. Csiki, Codrea, et al. 2010. See also Csiki et al. 2007.
20. Lydekker 1877.
21. Wilson and Upchurch 2003.
22. Casanovas et al. 1987; Lapparent 1947; Le Loeuff 1993; Lucas and Sullivan 2000; Wilson 2002, 2005.
23. Curry Rogers 2005; Powell 1992.
24. For example, see Curry-Rogers and Forster 2001.
25. Jianu and Weishampel 1999; Le Loeuff 1993.
26. Csiki and Grigorescu 2006a.
27. Csiki 1999; Dodson et al. 1998; Le Loeuff 1993, 2005; Le Loeuff et al. 1994; Powell 1992.
28. Witmer 2001.
29. Le Loeuff 1992; Powell 1992.
30. Stevens and Parrish 1999; but also see Taylor et al. 2009 for recent counter arguments.
31. Allain and Pereda-Suberbiola 2003; Chiappe 1995; Chiappe and Witmer 2002; Padian and Chiappe 1998; Witmer 1991.
32. Allain and Pereda-Suberbiola 2003; Buffetaut et al. 1986, 1988, 1996; Le Loeuff and Buffetaut 1998.
33. Csiki and Grigorescu 1998.
34. Nopcsa 1902a.
35. Andrews 1913.
36. Lambrecht 1929.
37. Harrison and Walker 1975.
38. Grigorescu and Kessler 1980.
39. Zoltán Csiki, pers. comm.
40. Brodkorb 1978.
41. Elzanowski 1983; Grigorescu 1984; Hope 2002; Makovicky and Norell 2004; L. Martin 1983; Olson 1985; Osmólska and Barsbold 1990.
42. Weishampel and Jianu 1996.
43. Sues 1978.
44. Csiki and Grigorescu 1998.
45. Codrea, Smith, et al. 2002; T. Smith et al. 2002.
46. Barsbold 1982; Barsbold and Osmólska 1999; Norell and Makovicky 2004; Ostrom 1969, 1990.
47. Naish and Dyke 2004.
48. Csiki and Grigorescu 1998.
49. T. Smith et al. 2002.

50. Chiappe et al. 2002.
51. Antunes and Sigogneau-Russell 1991; Pol et al. 1992; Rauhut and Zinke 1995; Sigé et al. 1997.
52. Csiki, Vremir, et al. 2010.
53. Carpenter 1997; Coombs and Maryańska 1990; Vickaryous et al. 2004.
54. Mantell 1832.
55. Bunzel 1871; Pereda-Suberbiola and Galton 1994, 1997.
56. Seeley 1881.
57. Nopcsa 1915, 1929a.
58. Codrea et al. 2001, 2002, 2010.
59. Pereda-Suberbiola and Galton 1994.
60. Vickaryous et al. 2004.
61. Carpenter 1997.
62. Nopcsa 1926d, 1929a.
63. Thulborn 1982.
64. Nopcsa 1897.
65. Bunzel 1871; Seeley 1881.
66. Matheron 1869a, 1869b.
67. Winand Brinkmann, then at the Free University in Berlin, put this taxonomic conflict to rest. He petitioned the International Committee of Zoological Nomenclature (ICZN), that great body of biologists who arbitrate cases of taxonomic controversy, to decide between *Rhabdodon* and *Mochlodon* (Brinkmann 1986). They voted in favor of *Rhabdodon* (International Committee of Zoological Nomenclature 1988). See also Brinkmann 1988.
68. Weishampel et al. 2003.
69. Buffetaut and Le Loeuff 1991b; Garcia et al. 1999; Pincemaille 1997.
70. Codrea et al. 2001, 2002, 2010; Weishampel et al. 2003.
71. Weishampel et al. 2003.
72. Codrea and Godefroit 2008; Godefroit et al. 2009; Weishampel et al. 2003.
73. Codrea and Godefroit 2008.
74. Alexander 1976, 1989; Farlow 1981; Thulborn 1982.
75. Codrea and Dica 2005; Vremir and Codrea 2002.
76. Codrea et al. 2001, 2002; Dalla Vecchia 2009; Weishampel et al. 1993.
77. Alexander 1989; Anderson et al. 1985; Colbert 1962.
78. Norman 1980, 1986.
79. Norman 1984; Norman and Weishampel 1985; Weishampel 1984.
80. Lull and Wright 1942; Taquet 1976; Weishampel et al. 1993.
81. Weishampel et al. 1993.

82. Dalla Vecchia 2009; Horner et al. 2004; Prieto-Márquez et al. 2006; Weishampel and Horner 1990; You, Luo, et al. 2003.

83. Grigorescu 1992b; Grigorescu et al. 1990, 1994, 2010; Weishampel et al. 1991.

84. Codrea, Smith, et al. 2002. See also T. Smith et al. 2002.

85. Zoltán Csiki, pers. comm.; *contra* Grigorescu et al. 1990.

86. Hirsch 1994; Hirsch and Packard 1987; Mikhailov 1991; Mikhailov et al. 1996.

87. Hirsch and Quinn 1990; Horner 1984.

88. Chiappe et al. 1998.

89. Cousin et al. 1994.

90. Grigorescu et al. 1990.

91. Grigorescu and Csiki 2006; Weishampel et al. 1993.

92. Grigorescu and Csiki 2006; Grigorescu et al. 2010.

93. Grigorescu et al. 1994.

94. Chiappe and Dingus 2001; Chiappe et al. 1998, 2001.

95. Grigorescu et al. 1994.

96. Horner 1984; Horner and Gorman 1988; Horner and Makela 1979; Horner and Weishampel 1988.

97. Weishampel and Horner 1994.

CHAPTER THREE. *Pterosaurs, Crocs, and Mammals, Oh My*

1. Wellnhofer 1991.
2. Fritsch 1881.
3. Seeley 1881.
4. Hooley 1914; Seeley 1869.
5. Nopcsa 1914b.
6. Nopcsa 1926c.
7. The first specimen of *Ornithodesmus*, called *Ornithodesmus cluniculus*, was considered to be a bird by Seeley in 1887, but he reassigned it to Pterosauria in 1901 (Seeley 1887a, 1901). There it sat, mostly through inertia, until Stafford Howse and Andrew Milner from Birkbeck College, London, had another look at it in 1992. It was only then that they realized that *Ornithodesmus* was not a pterosaur, but rather an imperfectly known maniraptoran theropod (Howse and Milner 1993).
8. Jianu et al. 1997.
9. Ősi and Fözy 2008. Unfortunately, this study makes no comparisons with pterosaur notaria.

10. Weishampel et al. 1991.
11. Buffetaut et al. 2002.
12. Buffetaut et al. 2002.
13. Henderson, in press.
14. Cai and Wei 1994.
15. Grigorescu et al. 1999; J. Martin et al. 2006.
16. J. Martin et al. 2010.
17. Nopcsa 1915, 1928a.
18. Matheron 1869a. See also J. Martin and Buffetaut 2008, who renamed this taxon *Massaliasuchus affuvelensis*.
19. Nopcsa 1928b.
20. Delfino et al. 2008.
21. Meers 1999.
22. Buscalioni et al. 2001, based on Romanian and Spanish material this paper ascribed to *A. precedens*. However, this referral may be erroneous, because of the subsequent discovery of a nearly complete *A. precedens* skull from Transylvania (see note 19).
23. Delfino et al. 2008. See also J. Martin and Delfino 2010.
24. J. Martin et al. 2006.
25. Delfino, Martin, et al. 2008.
26. J. Martin et al. 2006.
27. Company et al. 2005.
28. J. Martin et al. 2006.
29. Ross and Garnett 1989.
30. Nopcsa 1902a.
31. Folie and Codrea 2005. See also Grigorescu et al. 1999.
32. Grigorescu et al. 1999.
33. Folie and Codrea 2005.
34. Codrea et al. 2001, 2002.
35. Nopcsa 1923c.
36. Gaffney and Meylan 1992; Joyce 2007.
37. Codrea et al. 2010.
38. Gheerbrant et al. 2000; Lapparent de Broin and Murelaga 1996, 1999; Lapparent de Broin et al. 2004.
39. Kielan-Jaworowska et al. 2004.
40. Grigorescu and Hahn 1987; Grigorescu et al. 1985; Rădulescu and Samson 1986, 1997.
41. Csiki et al. 2005.
42. Rădulescu and Samson 1996.
43. Codrea, Smith, et al. 2002.

44. Csiki and Grigorescu 2000; T. Smith et al. 2002.
45. Krause 1982.
46. Krause and Jenkins 1982.
47. Maio 1988; Kielan-Jaworowska and Gambaryan 1994; Kielan-Jaworowska et al. 2004.
48. Folie and Codrea 2005; Grigorescu et al. 1999; Venczel and Csiki 2003.
49. Folie and Codrea 2005; Grigorescu et al. 1999.
50. Folie and Codrea 2005; Grigorescu et al. 1999.
51. Gardner 2000.
52. McGowan and Evans 1995.
53. Grigorescu et al. 1999.
54. Grigorescu et al. 1985.

CHAPTER FOUR. *Living on the Edge*

1. Weishampel et al. 1991 previously reported that the island was ~7,500 square km in area; this was a typographic error.
2. Nopcsa 1914b. See also Nopcsa 1923b.
3. Nopcsa 1914b, 12.
4. Ehrenberg 1975.
5. Nopcsa 1915.
6. Nopcsa 1917a.
7. Nopcsa 1923b, 1924.
8. Greene 1983; Suess 1916.
9. Nopcsa 1905b.
10. Nopcsa 1905b. See also Antonescu et al. 1983; Dincă et al. 1971; Grigorescu 1992; Mamulea 1953; Pop et al. 1971.
11. Nopcsa 1923b.
12. Greene 1983; Le Grand 1988; Suess 1883–1901.
13. Wegner 1912.
14. Weishampel and Reif 1984.
15. Nopcsa 1927.
16. Runcorn 1963.
17. Dietz and Holden 1970.
18. A. Smith et al. 1981, 1994.
19. A. Ziegler et al. 1983. For the Transylvanian region, see Panaiotu and Panaiotu 2010; Pătraşcu and Panaiotu 1990; Pătraşcu et al. 1992, 1993, 1994.
20. Barron 1983; A. Ziegler et al. 1993.
21. Dercourt et al. 1993.
22. Dercourt et al. 1993.

23. Berza et al. 1998; Burchfiel 1980; Csontos and Vörös 2005; Horváth 1974; Huismans et al. 1997; Linzer et al. 1998; Schmid et al. 1998; Willingshofer 2000; Willingshofer et al. 2001.

24. Burchfiel 1980; Săndulescu 1975; Willingshofer 2000.

25. Apulia is named after the present-day area in Italy's boot heel, its southeasternmost region. Long the gateway to and from the East, it has been conquered by legions of foreign rulers. Today it is known for its vineyards and olive groves. Rhodope is named after one the mountain ranges in the Balkan Peninsula, extending approximately 200 km from southeastern Bulgaria to northeastern Greece. Its highest peak, Musala, rises to 2,925 m.

26. Moesia was an ancient district bordered by the Danube and the Black Sea, once inhabited by some of the Thracian people. It was conquered by Rome between 30 and 20 BCE and became a Roman province in 15 CE. It corresponds approximately to present-day Bulgaria and Serbia.

27. Burchfiel 1980; Minkovska et al. 2002; Schmid et al. 1998; Willingshofer 2000; P. Ziegler 1988; Zweigel et al. 1998.

28. Mindszenty et al. 1995.

29. Costea et al. 1978.

30. Hallam 1992; Haq et al. 1988.

31. P. Ziegler 1990.

32. Huismans et al. 1997.

33. Estimated from Dercourt et al. 1993; Stampfli and Borel 2002. See also Zoltán Csiki, pers. comm.; Panaiotu and Panaiotu 2002; Patraşcu and Panaiotu 1990; Willingshofer 2000.

34. A. Bush 1997. Ocean surface temperatures indicated here are actually those estimated for the mid-Cretaceous (approximately 100 million years ago) by Lloyd 1982.

35. A. Bush 1997.

36. According to Marcin Machalski of the Instytut Palaeobiologica, Polska Akademia Nauk, Warsaw, Poland (pers. comm.), marine vertebrate faunas of eastern Europe are known, but they are not well documented in the literature. He and his collaborators recently described new mosasaur material from Poland (Machalski et al. 2003). Previous studies of mosasaurs from eastern Europe include Nikolov and Westphal 1976 and Sulimski 1968. Marine crocodilians are suspected to have inhabited this region, although their record (a thoracosaurine crocodilian) comes from the Danian (earliest Tertiary) of Poland (Zarski et al. 1998). Bony fish and sharks, on the other hand, have never been formally described, although they are known to be in private collections in Poland.

37. Anastasiu and Ciobuca 1989; Grigorescu 1992a; Stilla 1985; Weishampel et al. 1991.

38. Bojar et al. 2005; Grigorescu 1992a; Nopcsa 1905b; Therrien 2004, 2005, 2006.

39. Bojar et al. 2005; Therrien 2004, 2005, 2006.

40. Anastasiu and Ciobuca 1989; Grigorescu 1992a; Stilla 1985; Weishampel et al. 1991.

41. Weishampel et al. 1991.

42. Weishampel et al. 1991.

43. Weishampel et al. 1991.

44. Grigorescu 1983.

45. Codrea and Dica 2005; Therrien 2005, 2006.

46. Codrea and Godefroit 2008.

47. Bojar et al. 2005; Petrescu and Duşa 1982; Therrien 2004, 2005, 2006; Van Itterbeeck et al. 2004.

48. Therrien 2004, 2005, 2006.

49. Antonescu 1973; Antonescu et al. 1983; Duşa 1974; Lindfors et al. 2010; Mărgărit and Mărgărit 1962; Petrescu and Duşa 1980, 1982; Petrescu and Huică 1972; Van Itterbeeck et al. 2004, 2005.

50. Mărgărit and Mărgărit 1962; Petrescu and Duşa 1980, 1982.

51. Antonescu et al. 1983; Ion et al. 1988; Mamulea 1953; Pana et al. 2001.

52. Csiki 2008.

53. Schlüter 1983.

54. Burchfiel 1980; Csiki and Grigorescu 2007; Dercourt et al. 1986, 1993; Jianu and Boekschoten 1999a, 1999b; Sanders 1998.

55. Stampfli and Borel 2002.

56. Csiki and Grigorescu 2007; Jianu and Boekschoten 1999a, 1999b.

57. The two English sites are dated as earliest Late Cretaceous (Cenomanian) and therefore predate the other European faunas of interest by as much as 25 million years (Huxley 1867; Newton 1892). Several locations in Czechia have yielded a Late Cretaceous dinosaur fauna. Recently, a well-preserved femur belonging to an iguanodontian has been reported from the early Late Cretaceous (late Cenomanian; Fejfar et al. 2005). However, the majority of the remains from Czechia are too poorly preserved for any sort of taxonomic determination (Fritsch and Bayer 1905). Only four taxa have been reported from the Late Cretaceous of Sweden. One belongs to an indeterminate theropod (Persson 1959), another possibly to a bird (Nessov 1993), the third to an as-yet undescribed ornithopod (Nessov and Yarkov 1993), and the fourth to an indeterminate leptoceratopsid ceratopsian (Lindgren et al. 2008). For general aspects of the terrestrial faunas from the Late Cretaceous of Europe, see Brinkmann 1988; Buffetaut and Le Loeuff 1991a; Le Loeuff 1991; Nessov 1995; Weishampel 1990; Weishampel et al. 2004.

58. Literature on the Late Cretaceous faunas from France is quite considerable. The following are some of the more important recent studies: Buffetaut et al. 1988, 1989, 1995, 1996, 1997; Buffetaut and Le Loeuff 1991b, 1997, 1998; Cheylan 1994; Garcia et al. 1999; Laurent et al. 1997; Le Loeuff 1992, 1995; Le Loeuff and Buffetaut 1990, 1998; Le Loeuff and Moreno 1992; Le Loeuff and Souillat 1997; Pereda-Suberbiola 1992; Vasse 1995.

59. Buffetaut et al. 1993.

60. Viany-Liaud et al. 1994.

61. Antunes and Sigogneau-Russell 1991, 1996; Galton 1996.

62. Bataller 1960; Astibia et al. 1990, 1999; Brinkmann 1984; Buscalioni et al. 1997; Casanovas et al. 1987, 1999a, 1999b; Casanovas-Cladellas et al. 1985, 1988; Company et al. 1998; Pol et al. 1992; Prieto-Márquez et al. 2000; Sander et al. 1998; Sanz et al. 1995, 1999.

63. Buffetaut 1979; Bunzel 1871; Pereda-Suberbiola and Galton 1992; Sachs and Hornung 2006; Seeley 1881; Wellnhofer 1980.

64. Sachs and Hornung 2006.

65. Seeley 1883; Buffetaut et al. 1985; Mulder 1984; Mulder et al. 1997; Seeley 1881; Weishampel et al. 2000.

66. Riabinin 1945.

67. Wellnhofer 1994.

68. Dalla Vecchia 1997, 2009.

69. Debeljak et al. 1999.

70. Ősi 2003, 2004, 2005; Ősi et al. 2003, 2005, 2007, 2010.

71. Godefroit and Motchurova-Dekova 2010.

72. Arkhangelsky and Averianov 2003.

73. Averianov and Yarkov 2004; Yarkov and Nessov 2000.

74. Weishampel et al. 2004.

75. Weishampel et al. 2004.

76. Jianu and Boekschoten 1999a, 1999b. See also Sanders 1998.

77. MacArthur and Wilson 1967, 3–4.

78. J. Brown 1971; Hairston 1951; Mohr 1943.

CHAPTER FIVE. *Little Giants and Big Dwarfs*

1. Buffetaut 1987; see also Weishampel and White 2003.

2. Abel 1914 had earlier suggested that Empedocles (492–432 BC) considered the massive bones discovered on his native island of Sicily to be the remains of a race of giants. Mayor 2000 was unable to confirm Abel's suggestion through either the latter's own citations or in classical sources, indicating instead that

perhaps the connection between Empedocles and fossil bones was no more than imaginative supposition on Abel's part.

3. Boccaccio 1350.
4. Ginsburg 1984.
5. Plot 1677. See also Weishampel and White 2003.
6. Plot 1676, 134 [italics in original].
7. Brookes 1763.
8. Robinet 1768.
9. Delair and Sarjeant 1975; Halstead 1970; Halstead and Sarjeant 1993.
10. Delair and Sarjeant 1975; Desmond 1979, 1982, 1989; Torrens 1997.
11. Buckland 1824. See also Benson 2010; Edmonds 1979; Rudwick 1985.
12. Mantell 1825. See also Dean 1999.
13. Owen 1842, 1894; Rüpke 1994.
14. Dean 1999; Torrens 1997.
15. Owen 1842, 103.
16. Weishampel and Young 1996.
17. Glut and Brett-Surman 1997.
18. Holtz 2004.
19. Sereno et al. 1996.
20. Coria and Salgado 1995.
21. Dodson 1990.
22. Gillette 1994.
23. Bonaparte and Coria 1993.
24. Cope 1896. See also J. Brown and Maurer 1986; Damuth 1993; Gould 1966; Rensch 1959; Stanley 1973.
25. Alberch 1982; Alberch et al. 1979; McKinney 1988; McNamara 1988, 1997.
26. Alberch 1982; Alberch et al. 1979; McKinney 1988; McKinney et al. 1990; McNamara 1988.
27. McNamara 1997.
28. Lamar 1997.
29. By "dwarfed dinosaur," neither Nopcsa nor we are suggesting that a particular stature is achieved—it's not a matter of measuring these creatures with a meter stick. There is no absolute threshold below which we would recognize a dinosaur (or any other organism) as a dwarf. Nor are we speaking of shrew-, squirrel-, or dog-sized dinosaurs, although such body sizes could be possible for a dwarfed dinosaur, providing it evolved from a larger-bodied ancestor. Here is the quintessence of our recognition of dwarfing: downsizing that is attributable to phylogeny.

30. Nopcsa 1914b, 1915.
31. Nopcsa 1915, 18.
32. Nopcsa 1917a.
33. Boekschoten and Sondaar 1972; Roth 1992; Sondaar 1977.
34. Godfrey and Sutherland 1996; Shea 1989. See also Gould 1977.
35. McKinney and McNamara 1991. See also Blackstone 1987; Emerson 1986; Klingenberg 1998; McKinney 1986; McNamara 1986.
36. Gould 1977.
37. Haeckel 1866.
38. Gould 1977.
39. Alberch et al. 1979.
40. Alberch et al. 1979; Gould 1977; McKinney and McNamara 1991; McNamara 1997.
41. L. Brown and Rockwood 1986; Gould 1974; McKinney 1984; Morey 1994.
42. McKinney and Schoch 1985.
43. Shubin and Alberch 1986.
44. McNamara and Trewin 1993.
45. Wayne 1986.
46. Gould 1977.
47. Adams 1998; Adams and Pederson 1994.
48. See Horner et al. 2005; Jones and Gould 1999; Padian and Horner 2004; Padian et al. 2004.
49. Castanet 1987; Castanet et al. 1993; Chinsamy 1995; Chinsamy and Hillenius 2004; Chinsamy-Turan 2005; Curry 1999; Erickson and Tumanova 2000; Erickson et al. 2004; Horner et al. 1999, 2000, 2001; Klein and Sander 2008; Padian et al. 2004; Sander 2000. Nopcsa's interest in the histology of fossil bone, cut short by his suicide in 1933, can be seen in Nopcsa and Heidsieck 1933.
50. Varricchio 1997.
51. Erickson 2005; Erickson and Tumanova 2000; Erickson et al. 2006, 2009; Sander 1999, 2000; Sander and Clauss 2008; Sander et al. 2004, 2006.
52. Redelstorff et al. 2009.
53. Stein et al. 2010.
54. Redelstorff et al. 2009.
55. Long and McNamara 1995; Weishampel and Horner 1994.
56. Jianu and Weishampel 1999; Jianu et al. 1997; Weishampel et al. 1993.
57. Brooks and McLennan 1991; Farris 1970; Weishampel and Jianu 1997; Wiley et al. 1991.
58. Nelson and Platnick 1981; Nelson and Rosen 1981; Weishampel 1995; Witmer 1995.

59. D. Maddison and W. Maddison, 2000, *MacClade 4: Analysis of Phylogeny and Character Evolution*, Sinauer Associates, Sunderland, MA; K. C. Nixon, 1999, *Winclada*, www.cladistics.com/about_winc.htm; D. L. Swofford, 2002, *PAUP*: Phylogenetic Analysis Using Parsimony (and Other Methods)*, Sinauer Associates, Sunderland, MA.

60. Nopcsa 1900.

61. For a study of the replacement patterns of these kinds of teeth using theoretical morphology (introduced in chapter 4), see Weishampel 1991.

62. Weishampel et al. 1993.

63. Weishampel et al. 1993.

64. Alexander 1985, 1989, 1997.

65. Nopcsa 1915. Recently, Le Loeuff 2002 argued that these body size estimates for *Magyarosaurus dacus* are too low, based on material from this titanosaur in the collections of the Magyar Földtani Állami Intézet in Budapest and the Natural History Museum in London that suggests that adults reached a length of 10–15 m. We were aware of this material, but we did not include it because it was too incomplete to yield data for our study. In addition, these larger individuals are very rare in the total sample of specimens, and we regarded this rarity as representing very old individuals of continuously growing titanosaurs. Le Loeuff's alternative hypothesis was to identify our sample as young individuals, which necessitates a taphonomic explanation in which preservation biases favor juveniles over adults, perhaps coupled with community segregation into age cohorts. Whereas this is a very interesting hypothesis, histological evidence indicates that reproductively mature adults of *M. dacus* were small, and that this titanosaur was a paedomorphic dwarf. Csiki et al. 2007, 2010 resolved this quandary by recognizing two different titanosaur sauropods from the Haţeg Basin, the smaller, more common *M. dacus* and a larger, rarer form named *Paludititan nalatzensis*.

66. Curry Rogers and Forster 2001; Upchurch 1995, 1998; Wilson 2002; Wilson and Sereno 1998.

67. Jianu and Weishampel 1999.

68. The adult sample is significantly different from that of the growth series at $p = 0.01$, using the nonparametric quick test from Tsutakawa and Hewett 1977, which can be used to compare two populations represented by bivariate data, but with small sample sizes.

69. Again, employing the quick test from Tsutakawa and Hewett 1977, the *Magyarosaurus* sample is significantly different from the adult sample at $p = 0.03$, but not significantly different from the growth series. Unfortunately, we were unable to use the one large humerus identified by Le Loeuff 2002, because it is far from complete (and it may well be *Paludititan nalatzensis*).

70. This kind of mapping of features onto an existing cladogram is technically known as character optimization. For further information, see Brooks and McLennan 1991; Farris 1970; Wiley et al. 1991.

71. Curry Rogers and Forster 2001.

72. Weishampel and Jianu, unpublished manuscript; Weishampel et al. 2003.

73. Tsutakawa and Hewett 1977.

74. Buffetaut et al. 2003; Wellnhofer 1991.

75. Nopcsa 1926a.

76. Schindewolf 1993.

77. Schindewolf 1993, 300.

78. Alexander 1968; Burness et al. 2001; Calder 1984; Kleiber 1961; Lindstedt and Calder 1981; Pedley 1977; Schmidt-Nielsen 1972, 1975, 1979.

79. J. Brown and Maurer 1986.

80. Carlquist 1974; Case 1978; Foster 1964; Sondaar 1977; Van Valen 1973b.

81. Van Valen 1973b, 35. See also Benton et al. 2010; Case 1978; Sondaar 1977.

82. Roth 1992. See also Benton et al. 2010.

83. Lomolino 1984.

84. Csiki and Grigorescu 1998.

85. Hooijer 1957.

86. Goldman 2000.

87. J. Brown 1975; Grant 1965; Heany 1978; Lomolino 1985, 1986.

88. Damuth 1993; Maiorana 1973; Van Valen 1973b.

89. Gould 1977, 290.

CHAPTER SIX. *Living Fossils and Their Ghosts*

1. Courtenay-Latimer 1979, 7; Erdmann et al. 1998; Forey 1988, 1998; J. Smith 1956; Thomson 1991; Weinberg 2000.

2. For further information on the history of the discovery of coelacanths, see Pouyaud et al. 1999; Thomson 1991; Weinberg 2000.

3. Eldredge and Stanley 1984.

4. Darwin 1859, 105–108.

5. Delamare-Deboutteville and Botosancanu 1970. See also Schopf 1984.

6. Simpson 1953, 331.

7. Raup and Marshall 1980; Simpson 1944, 1953; Van Valen 1973a.

8. B. Schaeffer 1952; Westoll 1949.

9. Hennig 1965; Norell 1992.

10. Norell 1992. See also Benton 1990; Benton and Storrs 1994; Gauthier et al. 1988; Norell and Novacek 1992a, 1992b; Novacek and Norell 1982; Sereno

1991; Weishampel 1996; Weishampel and Heinrich 1992; Weishampel et al. 1993.

11. Benton and Storrs 1994. Two other techniques—the Stratigraphic Consistency Index, or SCI (Huelsenbeck 1994), and the Manhattan Stratigraphic Measure, or MSM (Siddall 1998)—have also been used in examining the fit between phylogeny and stratigraphy, but they are not as amenable to calculating the rates of character change that are found in this chapter.

12. Weishampel 1996.

13. Nopcsa 1923b.

14. Godefroit et al. 1998; Head 1998; Horner et al. 2004; Kirkland 1998; Norman 2004; Norman et al. 2004.

15. The age of the oldest member of the remaining hadrosaurids, a not particularly well-known euhadrosaurian named *Trachodon cantabrigiensis* from England (Lydekker 1888), is latest Early Cretaceous (i.e., late Albian).

16. Miller 1934, 2.

17. Weishampel et al. 1993.

18. Sereno et al. 1999. Pachycephalosaurs were not included in the comparisons made here, because their character data are few and almost completely restricted to the skull, thereby unnaturally depressing the rates of character change.

19. Cloutier 1991.

CHAPTER SEVEN. *Transylvania, the Land of Contingency*

1. Basinger 1990.

2. Gould 1989. See also Beatty 1995; Carrier 1995; Schaffner 1995.

3. Christie 1999; Matthews 1998.

4. Bush 1975; Coyne 1992; Endler 1977; Giddings et al. 1997; Kaneshiro 1989; Otte and Endler 1989.

5. Barigozzi 1982.

6. Barton and Charlesworth (1984) argued that during the development of a population by a few founders, the depletion of genetic variation and the effects of genetic drift would not be very great, unless the population remains small for a considerable number of generations.

7. Barton and Charlesworth 1984; Mayr 1942.

8. Mayr 1963.

9. Haldane 1956, 1957.

10. Mayr 1965.

11. Nopcsa 1923b. For a recent assessment of the biogeography of the Haţeg

vertebrate fauna, based on relationships within Europe, see Csiki 1997; Csiki and Grigorescu 2001; Weishampel et al. 2004.

12. Data from Weishampel et al. 2004.
13. This approach is described in detail in Weishampel and Jianu 1997.
14. The oldest euhadrosaurian comes from the late Albian of England (Lydekker 1888).
15. Godefroit et al. 2003; Weishampel et al. 2004; You, Luo, et al. 2003.
16. Head 1998; Horner et al. 2004; Kirkland 1998.
17. Buffetaut et al. 2001; Dalla Vecchia 2001; Dal Sasso 2003.
18. Dalla Vecchia 2009.
19. *Sensu* Horner et al. 2004.
20. Casanovas 1992; Casanovas et al. 1987; Casanovas-Cladellas and Santafé-Llopis 1993; López-Martínez et al. 2001.
21. Casanovas et al. 1999a, 1993; Dalla Vecchia 2006; Head 2001; Prieto-Márquez and Wagner 2009.
22. Dalla Vecchia 2006; Prieto-Márquez 2009.
23. Prieto-Márquez and Wagner 2009.
24. Head 2001; Pereda-Suberbiola et al. 2009; Prieto-Márquez et al. 2006.
25. Casanovas et al. 1999b; Prieto-Márquez and Wagner 2009.
26. Casanovas et al. 1999b; Dalla Vecchia 2006.
27. Casanovas et al. 1999b; Dalla Vecchia 2006; Pereda-Suberbiola et al. 2009.
28. See also Buffetaut et al. 1991.
29. Astibia et al. 1999.
30. Astibia et al. 1999; Sanz et al. 1999.
31. The actual situation may not be so straightforward. A very interesting occurrence—a yet-to-be-described euornithopod from the Late Cretaceous of Antarctica (Hooker et al. 1991; Angela Milner, pers. comm.; Milner et al. 1992) appears to have several dental and postcranial features in common with rhabdodontids and, if it proves to be a member of this clade, could greatly alter our biogeographic interpretations here.
32. Ősi 2005; Ősi and Makádi 2009; Ősi et al. 2003.
33. Le Loeuff 1995, and pers. comm.
34. Wilson and Upchurch 2003.
35. Delfino et al. 2008.
36. Ősi et al. 2007.
37. Weishampel et al. 2004.
38. Hirayama et al. 2000.
39. Danilov and Parham 2008; Joyce 2007.
40. Csiki and Grigorescu 2001, 2006; Gheerbrant et al. 2000; Peláez-Campomanes et al. 2000; Vianey-Liaud 1979, 1986.

41. Csiki and Grigorescu 2006b; Weishampel et al. 2004.

42. Titanosaur phylogeny appears to be in the forefront for establishing sauropod relationships (see Curry 2001; Curry Rogers and Forster 2001; Salgado et al. 1997; Sanz et al. 1999; Upchurch 1998; Upchurch et al. 2004; Wilson 2002; Wilson and Upchurch 2003), and efforts to put *Magyarosaurus* into this mix are now underway.

43. Baraboshkin et al. 2003.

44. Buffetaut et al. 1981; Le Loeuff and Buffetaut 1995.

45. Hanken and Wake 1993 and papers cited therein.

46. Lindstedt and Boyce 1985.

47. Boekschoten and Sondaar 1966; J. Brown 1975; Hutchinson 1959; Marshall and Corruccini 1978.

48. Azzaroli 1982; Valverde 1964.

49. Blueweiss et al. 1978; Calder 1984; Peters 1983; Stearns 1983, 1992.

50. Roth 1992.

51. Gould 1977; Roth 1992.

52. Gould and Vrba 1982; Vrba and Gould 1986.

53. Boekschoten and Sondaar 1966; Gould 1975, 1977; Hooijer 1975.

54. Norman and Weishampel 1985; Ostrom 1961, 1964, 1980; Weishampel and Norman 1989; Wing and Tiffney 1987.

55. Bakker 1978, 1986; Norman and Weishampel 1985, 1987, 1991; Weishampel 1985; Weishampel and Norman 1987, 1989.

56. Demment and Van Soest 1985; Janis 1976; Jarman 1974.

57. Demment and Van Soest 1985; Jarman 1974.

58. The monophyly of Euhadrosauria ensures a single migration from Europe.

59. Weishampel 1984.

60. Horner 1999, 2000; Horner and Currie 1994; Horner and Dobb 1997; Horner and Gorman 1988. See also Horner and Weishampel 1988; Weishampel and Horner 1994.

61. Gould 1977; MacArthur and Wilson 1967; McNaughton 1975; Pianka 1970, 1972; Southwood et al. 1974; Stearns 1976, 1983, 1992.

62. Bellemain and Ricklefs 2008.

63. Hoffman and Reif 1988, 1990.

CHAPTER EIGHT. *Alice and the End*

1. Van Valen 1973a.

2. Bell 1982; Lively 1996.

3. Buckling and Rainey 2002; Dietl 2003; Lovaszi and Moskat 2004.

4. Howard and Lively 2002; Jokela et al. 2000; Kawecki 1998; Martens and

Schön 2000; Ridley 1994. For an interesting about-face on the Red Queen hypothesis in interpretations of coevolutionary arms races, see Bergstrom and Lachmann 2003.

5. Barnosky 2001.
6. Futuyma 1998.
7. Tomkins 1996, 444.
8. Tzara 1992, 4.

GLOSSARY

Organismal groups that appear in this glossary are cladistically defined as either stem-based or node-based taxa and diagnosed on the basis of their shared derived characters (synapomorphies). See de Queiroz and Gauthier 1990 for the rationale for this decision. Also included in some of these entries is a reference to a recent comprehensive study of the taxon.

Acromion: A bony spine or process on the outer side of the scapula, to which some of the shoulder musculature is attached. This feature is very prominent in nodosaurid ankylosaurs.

Actinistia: The fleshy-finned fish, cladistically defined as the common ancestor of *Miguashaia* and *Latimeria*, and all the descendants of this common ancestor. This clade can be diagnosed by modifications of the appendicular fin skeleton, among other features. The most famous actinistian is the living coelacanth, *Latimeria*. See also "Coelacanth."

Albian: The interval of geologic time from approximately 112 to 99 million years ago.

Allopatry: When two species do not occur together, but exclude each other geographically.

Alpine Tethyan Sea: An oceanic basin, located where the western Alps presently are, that opened in the Early Jurassic (following the opening of the Atlantic Ocean) and closed during the Tertiary.

Altriciality: The kind of vertebrate ontogeny that is characterized by a short gestation and the birth of relatively underdeveloped, helpless young.

Alvarezsauridae: A node-based taxon defined as the most recent ancestor of *Mononykus* and *Alvarezsaurus*, and all the descendants of this ancestor. This clade is diagnosed by laterally compressed synsacral vertebrae, extreme modification of the forelimbs, and the absence of distal fusion in the pubis and ischium, among other characters. The phylogenetic position of alvarezsaurids is quite controversial. These theropods have been placed on either side of *Archaeopteryx*, making them either a true avialan clade or a very close outgroup. On the other hand, alvarezsaurids have been identified as the sister group to ornithomimosaurs. Other alvarezsaurids include *Shuvuuia*, *Mononykus*, and *Patagonykus* (Chiappe et al. 2002).

Ammonites: An extinct group of marine cephalopods whose closest living relative is the modern pearly nautilus. Their shells generally are in the form of flat spirals, although others uncoil to produce a straight conical shell. Known worldwide from the Silurian to the end of the Cretaceous, ammonites were among the top predatory invertebrates of their time.

Ankylosauria: The "armored" dinosaurs, a stem-based taxon defined as all eurypodan thyreophorans closer to *Ankylosaurus* than *Stegosaurus*. This clade can be diagnosed by numerous features of the skull, vertebral column, pelvis, and armor. Other ankylosaurs include *Edmontonia*, *Ankylosaurus*, and *Pinacosaurus* (Vickaryous et al. 2004).

Ankylosauridae: A stem-based taxon defined as those ankylosaurs more closely related to *Ankylosaurus* than to *Edmontonia*, and diagnosed by having a pyramidal squamosal boss, raised nuchal sculpturing, a premaxillary palate that is wider than long, a premaxillary notch, a deltoid quadratojugal boss, a postocular shelf, and a tail club, among other features. Other ankylosaurids include *Gobisaurus*, *Saichania*, and *Euoplocephalus* (Vickaryous et al. 2004).

Apparent polar wandering: The apparent migration, over Earth's surface, of Earth's magnetic poles through geologic time. Known from the direction of magnetization of many rocks, apparent polar wandering attests to the movement of continental landmasses and makes it possible to infer the relative movement of various continental blocks over different intervals of geologic time.

Aptian: The interval of geologic time from approximately 121 to 112 million years ago.

Apulia: A region in southeastern Italy, on the Adriatic coast and the Gulf of Taranto. In terms of plate tectonics, it is a microcontinental plate that once formed the promontory of northern Africa, but it detached to eventually suture with several other microcontinents and the Eurasian plate to form Italy, the inner region of Transylvania, and the Banat. See also "Moesia" and "Rhodope."

Aves: The monophyletic clade of all living birds; crown-group birds.

Avetheropoda: A node-based taxon consisting of *Allosaurus fragilis*, *Passer domesticus* (the living English sparrow), their most recent common ancestor, and all of its descendants, diagnosed by broad contact between the quadratojugal and the squamosal, palatine recesses, laterally displaced zygapophyses of the cervical vertebrae, a large and narrow iliac preacetabular fossa, an obturator process that is separated from the pubic plate, and the development of an accessory trochanter on the proximal lateral surface of the femur, among other characters. Other avetheropods include *Sinraptor*, *Carcharodontosaurus*, *Archaeopteryx*, and modern birds (Holtz and Osmólska 2004).

Avialae: A stem group encompassing living birds and all maniraptorans closer to them than to *Deinonychus*, a dromaeosaurid. This clade is diagnosed by many features, among them long, narrow, and pointed premaxillae; a quadrate that articulates with the prootic and the squamosal; unserrated teeth that are reduced in size and number; a pronounced acromion process on the scapula; a coracoid with a pronounced sternal process; forelimbs that are nearly as long as or longer than the hind limbs; forearms approximately as long as or longer than the humerus; and modifications of the foot, among other characters. This clade includes *Archaeopteryx*, *Hesperornis*, and Aves (Padian et al. 2004).

Azhdarchidae: A stem-based taxon defined as all pterosaurs more closely related to *Quetzalcoatlus* than to *Tapejara*, diagnosed by an elongation of the middle cervical vertebrae and the reduction or loss of the neural spines on these cervicals. In addition to *Quetzalcoatlus*, Azhdarchidae presently includes *Azhdarcho* (Kellner 2003).

Biogenic Law: Haeckel's term for his principle that ontogeny recapitulates phylogeny.

Biogeography: The study of the geographic distribution of life.

Biological constraints: The limits or boundary conditions on biological traits whose success or failure depends on alternatives in other organisms with which a species competes. These biological constraints are historical and developmental in their essence, since the ultimate arbiter is competitive advantage, and thus survival is imparted by design.

Biological determinism: A theory that claims that the traits of organisms are entirely determined by the interaction of genetic variation and the overarching control of natural selection.

Bioturbation: The biological activities that occur at or near the sedimental surface that cause the sediment to become mixed. Examples of these activities include burrowing and boring.

Campanian: The interval of geologic time from approximately 83.5 to 71.3 million years ago.

Casichelydia: A stem-based taxon defined as all testudines closer to *Crysemys picta* (the living painted turtle) than to *Proganochelys*. This clade is diagnosed by the loss of the lacrimal bone and duct, and a modification of the palate and braincase. Other casicheyids include *Kayentachelys* and two living turtles, the yellow-spotted Amazon River turtle (*Podocnemis unifilis*) and the eastern box turtle (*Terrapene carolina*) (Gaffney and Meylan 1988).

Cenomanian: The interval of geologic time from approximately 99 to 93.5 million years ago.

Ceratopsia: A stem clade defined as all of Marginocephalia closer to *Triceratops* than to *Pachycephalosaurus*, which is diagnosed by a high external naris separated from the ventral border of the premaxilla by a flat area, a rostral bone, an enlarged premaxilla, well-developed lateral flaring of the jugal, wide dorsoventral length of the infraorbital ramus of the jugal, and contact of the palatal extensions of the maxillae rostral to the choana. Other ceratopsians include *Psittacosaurus*, *Protoceratops*, and *Styracosaurus* (You and Dodson 2004).

Ceratosauria: A stem-based taxon defined as those theropods more closely related to *Ceratosaurus nasicornis* than to birds. This clade can be diagnosed by modifications of the vertebral column and pelvis, among others. Other ceratosaurians include *Coelophysis*, *Syntarsus*, and *Carnotaurus*. There is some question about the monophyly of this group (Tykoski and Rowe 2004).

Cimmerian block: An elongate and narrow continental block that, when it began its northern motion approximately 280 million years ago, extended from modern-day Gibraltar to the eastern margin of Australia, where it formed the southern margin of the Paleotethyan Ocean. The Cimmerian block docked with the Eurasian continental landmass, thereby creating the largest extent of the Neotethyan Ocean, about 200 million years ago.

Clade: A group of biological taxa that includes all the descendants of a single common ancestor.

Cladistics: A system of biological taxonomy that defines taxa by shared unique characteristics not found in ancestral groups and uses inferred evolutionary relationships to arrange taxa in a branching hierarchy, such that all members of a given taxon have the same ancestors. See also "Phylogenetic systematics."

Cladogram: A branching diagrammatic tree used in cladistic classification to illustrate phylogenetic relationships.

Coelacanth: Usually refers to the living form, *Latimeria*, and its two species (*L. chalumnae* and *L. menadoensis*), as well as their closest (but extinct) relatives. See also "Actinistia."

Coevolution: Evolution involving successive changes in two or more ecologically interdependent species that affect their interactions.

Comparative anatomy: The study of the anatomy of several groups of organisms.

Cope's Rule: The tendency for size increases to occur in evolutionary lineages.

Court Jester hypothesis: A model of extinction in which changes in the physical environment are the initiators of major changes in organisms.

Crocodylia: A node-based (crown-group) taxon defined as the common ancestor of the living *Alligator mississippiensis* and the living *Gavialis gangeticus*, and all the descendants of this common ancestor. This clade is diagnosed by the construction of the skull roof and the mandible. Modern crocodilians include alligators, crocodiles, gavials, and caimans, among others (Buscalioni et al. 2001).

Crown group: The clade made up of all living members of the group, the ancestor of that clade, and all the descendants of that ancestor. Examples include Mammalia, Aves, and Archosauria.

Cryptodira: The hidden-necked turtles, a node-based taxon defined as the common ancestor of *Kayentachelys* and *Crysemys picta* (the living painted turtle), and all the descendants of this common ancestor. This clade can be diagnosed by several features of the palate, the jaw system, and the skull roof, as well as the turtle's unique way of folding its neck. Other cryptodires include *Meiolania*, living soft-shelled freshwater turtles, and the modern Galápagos giant tortoise (*Geochelone nigra*) (Gaffney and Meylan 1988).

Dacians: The people who inhabited the region known as Dacia (in what is now Romania, Moldova, and parts of Serbia, Bulgaria, Hungary, and

Ukraine) between the second century BCE and the second century CE, when they were conquered by the Romans.

Definition: A statement about the membership of a clade, based on common ancestry.

Dental battery: Upper and lower dentition, composed of abundant (up to 60 tooth positions), closely packed cheek teeth and consisting of one to three functional teeth and up to five replacement teeth per tooth position. Dental batteries are known in hadrosaurid dinosaurs.

Diagnosis: The list of features that are synapomorphies, delimiting the taxon that has been defined.

Diastema: A space between teeth in a jaw, usually referring to the gap between the front of the jaws and the cheek teeth (as seen in hadrosaurids), or the gap between the incisors and premolars (in horses).

Dinosauria: A node-based taxon consisting of *Triceratops*, Neornithes, their most recent common ancestor, and all their descendants. It is diagnosed by the loss of the postfrontal, elongated deltopectoral crest on the humerus; a brevis shelf on the ventral surface of the postacetabular part of ilium; an extensively perforated acetabulum; a tibia with a transversely expanded, subrectangular distal end, as well as a caudolateral flange and a depression for the astragalus; and an ascending astragalar process on the cranial face of the tibia (Benton 2004).

Dromaeosauridae: A node-based taxon defined as all the descendants of the most recent common ancestor of *Microraptor zhaoianus*, *Sinornithosaurus milleni*, and *Velociraptor mongoliensis*. This clade can be diagnosed by features of the skull roof, the vertebrae, and a sickle-shaped claw on the second digit of the foot, among others. Other dromaeosaurids include *Saurornitholestes* and *Deinonychus* (Norell and Makovicky 2004).

Dsungaripteroidea: A node-based pterosaur taxon defined as the most recent common ancestor of *Nyctosaurus* and *Quetzalcoatlus*, and all of its descendants, diagnosed by edentulous jaws, a notarium, and a pneumatic foramen on the proximal humerus, among other features. Other dsungaripteroids include *Pteranodon*, *Noripterus*, and *Anhanguera* (Kellner 2003).

Euhadrosauria: The true hadrosaurids, a node-based taxon defined as the common ancestor of *Edmontosaurus* and *Corythosaurus*, and all the descendants of this common ancestor. The clade is diagnosed by modifications of the jaws. Other euhadrosaurs include *Gryposaurus*, *Parasaurolophus*, and *Charonosaurus* (Horner et al. 2004).

Eusuchia: The true suchians, a node-based taxon defined as the common ancestor of *Hylaeochampsa* and *Alligator*, and all the descendants of this common ancestor. This clade can be diagnosed by modifications of the palate and the vertebrae. Other eusuchians include *Alligator*, *Caiman*, and *Gavialis* (Buscalioni et al. 2001).

Foraminifera: A group of large, chiefly marine protozoans—usually having calcareous shells that often are perforated with minute holes for the protrusion of slender pseudopodia—that form the bulk of chalk and nummulitic limestone.

Founder effect: The principle that the founders of a new, isolated population carry a gene pool that is not representative of that of the population as a whole.

Functional morphology: The study of how biological structures work, which, in paleontology, entails the use of analogies with mechanical and extant animal models.

Ghost lineage: The lineage of an organism for which there is no physical record, but whose existence can be inferred by phylogeny calibrated by stratigraphy. The length of time incorporated into a ghost lineage is known as its ghost lineage duration (GLD).

Gondwana: The region consisting of the once-connected Indian subcontinent and the landmasses of the Southern Hemisphere.

Hadrosauridae: The duck-billed dinosaurs, a node-based taxon defined as the common ancestor of *Telmatosaurus* and *Corythosaurus*, and all the descendants of this common ancestor. This clade can be diagnosed by numerous features involving dentition, the skull and palate, the vertebral column, and the appendicular skeleton. Other hadrosaurids include *Gryposaurus*, *Edmontosaurus*, and *Parasaurolophus* (Horner et al. 2004).

Hadrosaurinae: The non-lambeosaurine euhadrosaurians, a stem-based taxon defined as all euhadrosaurs more closely related to *Edmontosaurus* than to *Corythosaurus*. This clade can be diagnosed on the basis of modifications of the nasal region of the skull. Other hadrosaurines include *Gryposaurus*, *Maiasaura*, and *Saurolophus* (Horner et al. 2004).

Hadrosauroidea: A stem-based taxon defined as all iguanodontians more closely related to *Corythosaurus* than to *Iguanodon*. This clade is diagnosed by modifications of the facial skeleton, the tail vertebrae, the pelvis, and the hind foot. Other hadrosauroids include *Shuangmiaosaurus*, *Altirhinus*, and *Eolambia* (You et al. 2003).

Heterochrony: Evolutionary differences in the features of organisms due to ontogenetic changes in the relative rates or timing of the development of organismal traits. When a descendant is ontogenetically less well developed than its ancestor, the resulting morphology is termed paedomorphic. When its development goes beyond that of its ancestor, the resulting morphology is called peramorphic. See also "Paedomorphosis" and "Peramorphosis."

Hierarchy: In evolutionary biology, the ordering of organisms by a pattern of common descent; clades within clades.

Histology: The study of tissue structure or organization.

Historical contingency: The conditions by which the details of evolutionary history are unpredictable, that is, determined by chance events and/or by improbable events with large consequences.

Homology: A similarity between two organisms, due to the inheritance of the same feature from a common ancestor.

Homoplasy: A similarity between two organisms, due to separate and independent inheritances from different ancestors. Evolutionary convergence is one kind of homoplasy.

Iguanodontia: A stem-based taxon defined as all euornithopods closer to *Edmontosaurus* than to *Thescelosaurus*. This clade can be diagnosed by numerous modifications of the snout, the sternum, the pelvis, and the manus. Other iguanodontians include *Dryosaurus*, *Ouranosaurus*, and hadrosaurids (Norman 2004).

Insularity: As used in this volume, the state or condition of being cut off or isolated from other organisms.

Island Rule: The common pattern of a dwarfing of mammals evolving on islands.

***K* selection / strategy:** Selection on individuals in populations at or near the carrying capacity of their environments, usually favoring the production of a few, slowly developing young that are well adjusted in form and function to their (usually stable) environment. *K*-selected individuals are good competitors in conditions of density-dependent mortality.

Lambeosaurinae: The hollow-crested hadrosaurids, a stem-based taxon defined as all euhadrosaurs more closely related to *Corythosaurus* than to *Edmontosaurus*. This clade can be diagnosed by modifications of the jaw system and the skull roof. Other lambeosaurines include *Lambeosaurus*, *Corythosaurus*, and *Tsintaosaurus* (Horner et al. 2004).

Laurasia: The region consisting of the once-connected landmasses of the Northern Hemisphere, except for the Indian subcontinent.

Laws of physics: From an evolutionary perspective, the universal/ahistorical laws that dictate the general configuration of biological forms. These laws arise from the physical tenets of geometry, scaling laws, and packing principles.

Life-history strategies: Selected sets of adaptations to local environments, involving such quantitative aspects of life history as fecundity, the timing of maturation, and the frequency of reproduction. Responses to r and K selection represent two different life-history strategies.

Lithosphere: The solid, outer part of Earth's core, composed of rock essentially like that exposed at the surface and usually considered to be about 80 km thick.

Living fossil: An organism that has remained essentially unchanged from much earlier geologic times and whose close relatives are usually extinct. Living fossils are descendants of extinct organisms greatly removed in time from their sister group (i.e., having a large GLD) and with very little character transformation over that GLD.

Lycian Sea: A small oceanic basin, located along the southwestern coast of Anatolia in northern Turkey. The Lycian Sea opened in the late Early Cretaceous as a second basin adjoining that of the Vardar Sea, and it persisted into the Tertiary.

Maastrichtian: The interval of geologic time from approximately 71.3 to 65 million years ago.

Maniraptora: A stem-based clade consisting of *Passer domesticus* (the living English sparrow) and all taxa closer to it than to *Ornithomimus velox*. This clade is diagnosed by the presence of ossified sternal plates, changes in the elbow, a semilunate carpal, modification of the hand, changes in the pelvis (including the rearward rotation of the pubis), and broad pennaceous feathers on the forelimb and tail, among other characters. Other maniraptorans include troodontids, oviraptorosaurs, and dromaeosaurids (Holtz and Osmólska 2004).

Marsupialia: A node-based (crown-group) taxon defined as the most recent common ancestor of the modern American opossum (*Didelphis virginiana*) and the modern wombat (*Lasiorhinus latifrons*). This clade is diagnosed by modifications of the ear and the braincase; the number of incisors, premolars, and molars; and, of course, by their reproductive biology (paired female internal genitalia, a very short gestation period, and early development in a pouch). The stem-based taxon that contains Marsupialia is known as Metatheria.

Megaloolithidae: One of the categories in the classification of fossil eggs, in which the shell is formed from shell units of tabular calcite crystals arranged in a radiating pattern (dinosauroid-spherulitic basic type). Each shell unit forms an external bump or node (tubospherulitic structural morphotype).

Moesia: An ancient country and then a Roman province in southeastern Europe, presently represented by Serbia and Bulgaria (south of the Danube from the Drina to the Black Sea). In terms of plate tectonics, it is the microcontinental plate of the central Peri-Tethyan region that was the first to collide and then suture with the Eurasian continental plate. See also "Apulia" and "Rhodope."

Monophyly: The condition of having evolved from a single common ancestral form.

Monotremata: A node-based (crown-group) clade defined as the most recent common ancestor of the modern echidna (*Tachyglossus*) and the modern platypus (*Ornithorhynchus*), and all the descendants of this ancestor. This clade is diagnosed by a reduced or absent jugal, a slender dentary with only a vestige of the coronoid process, the absence of auditory bullae, and complex modifications of the shoulder girdle, among other features. The stem-based clade that contains Monotremata is known as Prototheria.

Morphospace: A multidimensional space comprising the set of all theoretically possible morphologies, based on geometric or growth parameters of particular groups of organisms.

Multituberculata: A node-based taxon, now extinct, defined as the most recent common ancestor of *Zofiabaatar* and *Kogaionon*, and all the descendants of this common ancestor. This clade can be diagnosed by modifications of the teeth, the skull, and the pelvis. Multituberculates were the most diverse and widespread of all mammals at the end of the Mesozoic and into the Tertiary (Kielan-Jaworowska et al. 2004).

Neo-Lamarckism: A late nineteenth- and early twentieth-century theory of evolutionary transformation, named after Jean-Baptiste de Monet de Lamarck (1744–1829), a French naturalist. This theory postulates that adaptations arise as characters are acquired by active organic responses to the environment, which are then passed on to offspring through heredity.

Neotethyan Ocean: The ocean that developed south of the northward-moving Cimmerian block sometime during the Permian. Similar to the earlier Tethyan Ocean, this younger ocean once again separated Laurasia and Gondwana and opened to the east. The Neotethyan Ocean was present throughout the Mesozoic.

Neoteny: The retention of juvenile features in adults, due to a slowing down of growth rates.

Nodosauridae: A stem-based clade defined as all ankylosaurs more closely related to *Edmontonia* than to *Ankylosaurus*. This clade can be diagnosed by features of the skull, the shoulder and pelvic girdles, and armor. Other nodosaurids include *Animatarx*, *Struthiosaurus*, and *Sauropelta* (Vickaryous et al. 2004).

Notarium: The fusion of the first six to eight dorsal vertebrae, a condition found only in some pterosaurs and many birds.

Ontogeny: The life course of the development of an individual organism; the history of an individual, both embryonic and postnatal.

Optimization: Establishing the most parsimonious sequences of character transformation on an existing cladogram.

Ornithischia: The bird-hipped dinosaurs, a stem-based taxon defined as all dinosaurs that are closer to *Triceratops* than they are to Saurischia. This clade can be diagnosed by the predentary bone capping the front of the lower jaws and a rearwardly rotated pubic bone, among other features (Benton 2004).

Ornithocheiridae: A poorly defined and diagnosed group of pterodactyloid pterosaurs, now regarded as a "wastebasket" taxon.

Ornithomimosauria: The ostrich-mimicking theropods, a node-based taxon consisting of the last common ancestor of the clade defined by *Ornithomimus edmontonicus* and *Pelecanimimus polyodon*, and all of its descendants. This clade is diagnosed by an inflated cultriform process forming a bulbous, hollow structure; a premaxilla with a long, tapering, subnarial ramus that separates the maxilla and the nasal for a distance, caudal to the naris; an elongated and subtriangular dentary; and a tightly bound distal radius and ulna, among numerous other characters. Other ornithomimosaurs include *Garudimimus*, *Struthiomimus*, and *Gallimimus* (Makovicky et al. 2004).

Ornithopoda: A stem-based taxon defined as all genasaurians (ornithischians with cheeks) more closely related to *Parasaurolophus walkeri* than to *Triceratops horridus*. This clade can be diagnosed by modifications of features of the jaw system and the back of the skull. Other ornithopods include *Hypsilophodon, Tenontosaurus,* and *Iguanodon* (Butler et al. 2008; Norman et al. 2004).

Oviraptoridae: A stem-based taxon consisting of the most inclusive oviraptorosaurian clade, containing *Oviraptor philoceratops* but not *Caenagnathus collinsi*, diagnosed by a narrow snout; a pneumatized premaxilla; a skull roof and quadrate, fused nasals; large and subquadrate infratemporal fenestra; and many other (principally cranial) modifications. Other oviratptorids include *Citipati, Ingenia,* and *Conchoraptor* (Osmólska et al. 2004).

Oviraptorosauria: A stem-based taxon defined as all maniraptorans closer to *Oviraptor philoceratops* than to *Passer domesticus* (the living English sparrow), diagnosed by a crenulated ventral margin of the premaxilla, a parietal that is at least as long as the frontal, an ascending process of the quadratojugal bordering more than three-quarters of the infratemporal fenestra, a U-shaped mandibular symphysis, an edentulous dentary, and a cranial process on the pubic foot that is longer than the caudal process. Other oviraptorosaurs include *Incisivosaurus, Chirostenotes,* and *Khaan* (Osmólska et al. 2004).

Pachycephalosauria: A stem-based clade defined as all taxa more closely related to *Pachycephalosaurus wyomingensis* than to *Triceratops horridus*, diagnosed by a thickened skull roof, the exclusion of the frontal from the orbital margin, tubercles on the caudolateral margin of the squamosal, a caudal basket of fusiform ossified tendons, a medial process on the iliac blade, and a pubis that is nearly excluded from the acetabulum, among other features. Other pachycephalosaurs include *Wannanosaurus*, *Homalocephale*, and *Stygimoloch* (Vickaryous et al. 2004).

Paedomorphosis: The retention of ancestral juvenile characters by later ontogenetic stages of descendants. Paedomorphosis is produced by neoteny (the deceleration of growth rate in descendants), progenesis (the early termination of the development of a feature in descendants), and postdisplacement (the late onset of the development of a feature in descendants). See also "Heterochrony" and "Peramorphosis."

Paleobiology: The branch of paleontology concerned with the biology of fossil organisms. This discipline originated with the work of Vladimir Kovalevsky (1842–1883); was taken up as a distinct discipline in the early twentieth century in the studies of Dollo, Abel, Nopcsa, Wiman, Versluys, and others; and was reintroduced as a major research field in the 1970s.

Paleogeography: The geographic disposition of continents and oceans over geologic time.

Paleosol: A fossil soil; a soil formed during an earlier period of pedogenesis.

Paleotethyan Ocean: A mainly Paleozoic ocean (Silurian–Jurassic) that separated the Laurasian and Gondwanan regions of Pangaea, opening to the east. Beginning in the Permian, the Cimmerian block moved northward, diminishing the size of the Paleotethyan Ocean, until the latter disappeared in the Early Jurassic, replaced by the Neotethyan Ocean.

Pangaea: A region believed to have been the once-connected landmasses of the Southern Hemisphere and the Northern Hemisphere, dating from the Mississippian–Pennsylvanian boundary (323 million years ago) to the transition between the Triassic and the Jurassic (206 million years ago).

Parapatry: The condition in which diverging populations occupy distinct but contiguous areas.

Paris Basin: The depositional basin surrounding the area of Paris, France, that preserves sedimentary rocks of Mesozoic and Tertiary age.

Parsimony: The principle of parsimony (also known as "Ockham's Razor") requires ad hoc assumptions to be minimized as far as possible in scientific explanations of natural phenomena. For cladistics, this means that, from the millions of theoretically possible cladograms, those should be preferred that minimize the number of necessary assumptions of nonhomology (homoplasies).

Peramorphosis: The extension of descendant ontogeny beyond that of the ancestral condition. Peramorphosis is produced by acceleration (an increase in the growth rate in descendants), hypermorphosis (the late termination of the development of a feature in descendants), and predisplacement (the early onset of the development of a feature in descendants). See also "Heterochrony" and "Paedomorphosis."

Peripatry: Occurring at the periphery.

Phylogenetic systematics: A methodology, described by Willi Hennig, for the reconstruction of phylogenetic trees and the discovery of monophyletic groups by the exclusive use of shared (homologous) derived character states. See also "Cladistics."

Phylogeny: The evolutionary history of the relationships of species within a clade.

Pindos Sea: A small oceanic basin, extending across present-day northern Greece, that opened in the Late Triassic and closed during the Late Cretaceous.

Placentalia: A node-based taxon defined as the most recent common ancestor of the modern two-toed tree sloth (*Cholepus didactylus*) and humans (*Homo sapiens*), and all of the descendants of this ancestor. This clade is diagnosed by a prolonged intrauterine gestation, the loss of epipubic bones, and changes in dentition, among other features. The stem-based clade that contains Placentalia is known as Eutheria.

Plate tectonics: The theory in geology in which Earth's lithosphere is divided into a small number of plates that float on, and travel independently over, the mantle. Much of Earth's seismic activity occurs at the boundaries of these plates.

Pleurodira: The side-necked turtles, a node-based taxon defined by the common ancestor of *Proterochersis* and Podocnemidae, and all the descendants of this common ancestor. This clade can be diagnosed by a unique pulley arrangement of the jaw musculature, a suturing of the pelvis into the shell, and particular aspects of the jaw system. Other pleurodires include the extinct *Stupendemys*, the living yellow-spotted

Amazon River turtle (*Podocnemis unifilis*), and the living mata mata (*Chelus fimbriatus*) (Gaffney and Meylan 1988).

Polish Trough: The depositional basin that extends from what is now the North Sea to the Black Sea.

Precociality: The kind of vertebrate ontogeny that is characterized by the birth of relatively mature, independent young.

Predentary: The single bone that caps the front of the paired mandibles, found solely in Ornithischia.

Pterodactyloidea: A node-based taxon defined as the most recent common ancestor of *Pterodactylus* and *Quetzalcoatlus*, and all of its descendants. This clade can be diagnosed by modifications of the facial skeleton, the vertebral column, the hand, and the foot. Other pterydactyloids include *Pteranodon, Ornithocheirus,* and *Azhdarcho* (Kellner 2003).

Pterosauria: The flying reptiles, a node-based taxon defined as the most recent common ancestor of *Eudimorphodon* and *Quetzalcoatlus*, and all the descendants of this common ancestor. This clade can be diagnosed by numerous features, including an enlargement and modification of the skull, a modification of the vertebral column, and an elongation and modification of the forelimb, among other characters. Other pterosaurs include *Dimorphodon, Rhamphorhynchus,* and *Anhanguera* (Kellner 2003).

***r* selection/strategy:** Selection on individuals in populations well below the carrying capacity of their environments, usually favoring the early and rapid production of large numbers of quickly developing young. *r* selection generally operates in ecological situations favoring a rapid increase in population size—either because environments fluctuate so severely and unpredictably that organisms do best by making as many offspring as quickly as possible; or because ephemeral, superabundant resources can best be utilized to build up the population size before the inevitable exhaustion of these resources.

Red Queen hypothesis: The theory that describes how the coevolution of competing species creates a dynamic equilibrium in which the probability of extinction remains fairly constant over time. As one species evolves improvements that make it more competitive, its competitors experience selection pressures that force them to evolve in order to keep pace with it.

Regression: In a geologic context, the retreat of the sea from land areas.

Rhabdodontidae: A node-based taxon defined as the common ancestor of *Rhabdodon priscus* and *Zalmoxes robustus*, and all the descendants of this common ancestor. The clade is diagnosed by abundant sharp ridges on the dentary and the maxillary tooth crowns, the straight dorsal margin of the ilium, and a distinctly bowed femur, among other features (Weishampel et al. 2003).

Rhamphotheca: The cornified beak of birds, ornithischians, and turtles, consisting of the keratinized covering of the tips of the upper and lower jaws.

Rhodope: Mountains in southern Bulgaria and northeastern Greece. In terms of plate tectonics, it is the microcontinental plate sandwiched between Moesia and Eurasia to the east and north and Apulia to the west. After the collision of Moesia with Eurasia, Rhodope then collided with Moesia, and it was subsequently struck by Apulia. Full suturing of these three microcontinents with Eurasia took place in the mid-Tertiary. See also "Apulia" and "Moesia."

Rudists: Sedentary clams (Bivalvia; Rudistacea) that have a superficially coral-like form and lifestyle, including their habit of gregarious reef building.

Santonian: The interval of geologic time from approximately 85.8 to 83.5 million years ago.

Saurischia: The lizard-hipped dinosaurs, a stem-based clade defined as all dinosaurs closer to birds than they are to Ornithischia. This clade can be diagnosed by modifications of the skull, the pelvis, and the hind limb (Benton 2004).

Sauropoda: The gigantic, long-necked, long-tailed dinosaurs, a stem-based clade defined as all sauropodomorphs more closely related to *Saltasaurus* than to *Plateosaurus*. Features diagnosing this clade include an extensive modification of the neural arch and the spine of the cervical vertebrae, relatively short forelimbs, changes in the forefoot, a compressed distal end of the ischial shaft, several modifications of the size and shape of the femur, and alterations of the hind foot, among other features. Other sauropods include *Apatosaurus*, *Diplodocus*, and *Brachiosaurus* (Upchurch et al. 2004).

Selmacryptodira: A stem-based taxon defined as all cryptodires more closely related to *Crysemys picta* (the living painted turtle) than to

Kayentachelys. This clade is diagnosed by palatal, middle ear, and braincase features, and by the loss of palatal teeth (Gaffney and Meylan 1988).

Severin Ocean: A small oceanic basin, located in southwestern Romania, that opened in the Late Jurassic and closed during the Late Cretaceous.

Skeletochronology: The use of bone histology to calibrate growth stages and determine growth rates.

Speciation: The process of biological species formation, usually regarded as the splitting of one species lineage into two lineages.

Stegosauria: A stem-based clade defined as all taxa more closely related to *Stegosaurus* than to Ankylosauria, diagnosed by a flattened dorsal surface of the parietals, dorsal neural arches that are at least 1.5 times as high as the dorsal centra, a prominent triceps tubercle and a descending ridge caudolateral to the deltopectoral crest on the humerus, fusion of the wrist bones in adults, numerous modifications of the hind foot, parasagittal rows of plates or spines, and the loss of ossified epaxial tendons, among other features. Other stegosaurs include *Huayangosaurus*, *Hesperosaurus*, and *Dacentrurus* (Galton and Upchurch 2004).

Stratigraphy: The part of geology that deals with the origin, composition, distribution, and succession of strata.

Sympatry: Occurring in the same geographic region.

Taphonomy: The study of the processes (i.e., burial, decay, and preservation) that affect animal and plant remains as they become fossilized.

Taxon (pl. taxa): A taxonomic group or entity.

Tectonics: A branch of geology concerned with the structure of Earth's crust.

Testudines: All turtles. This node-based (crown-group) taxon is defined as the most recent common ancestor of *Proganochelys* and *Crysemys picta* (the living painted turtle), and all the descendants of this ancestor. The clade is diagnosed by a bony shell consisting of a carapace and a plastron, edentulous jaws covered with a rhamphotheca, and the loss of several skull-roof bones, among other features (Gaffney and Meylan 1988).

Tetanurae: A stem-based clade that includes *Passer domesticus* (the living English sparrow) and all taxa sharing a more recent common ancestor with it than with *Ceratosaurus nasicornis*. This clade can be diagnosed by an axially reduced and rodlike neural spine on the second cervical vertebra (axis), a prominent acromion on the scapula, modifications of the hand and the wrist, and changes in femoral morphology, among other features. Other tetanurans include *Megalosaurus*, *Allosaurus*, *Tyrannosaurus*, dromaeosaurids, troodontids, and birds (Holtz et al. 2004).

Tethyan Ocean: The sea believed to have extended into eastern Pangaea, and later to have separated Laurasia to the north from Gondwana to the south. The Mediterranean is a remnant of it.

Tetrapoda: The vertebrates with legs, cladistically defined as the common ancestor of Amphibia and Amniota (mammals, turtles, lizards and snakes, crocodilians, and birds), and all the descendants of this common ancestor. This clade can be diagnosed by a host of modifications in the braincase, the ear, the pelvis, and the limb skeleton.

Theoretical morphology: The field of evolutionary biology that involves the mathematical simulation of organic morphogenesis and an analysis of the possible spectrum of organic form via hypothetical morphospace construction.

Theria: A node-based taxon defined as the most recent common ancestor of the modern American opossum (*Didelphis americana*) and humans (*Homo sapiens*), and all the descendants of this ancestor. This clade is diagnosed by the presence of a suprascapular fossa on the scapula, the absence of shelled eggs for reproduction, mammary glands, and a coiling of the cochlea in the inner ear, among other characters (Novacek et al. 1988).

Therizinosauroidea: A node-based taxon defined as the least inclusive clade containing *Therizinosaurus* and *Beipiaosaurus*, diagnosed by a toothless premaxilla with a sharp, ventrally projecting rim; a greatly elongated external naris; a palate with greatly elongated vomers and rostrally reduced pterygoids; a greatly enlarged and pneumatized basicranium; laterally compressed, symmetrical maxillary and dentary teeth with coarse serrations and cylindrical roots; modifications of the wrist and the hand; a deep and long preacetabular process of the ilium that flares outward at a right angle to the sagittal plane; and short and broad metatarsals, among other features. Other therizinosauroids include *Alxasaurus*, *Segnosaurus*, and *Nothronychus* (Clark et al. 2004).

Theropoda: The predatory dinosaurs, a stem-based clade defined as all saurischians more closely related to *Passer domesticus* (the living English sparrow) than to *Cetiosaurus oxoniensis*. Diagnosing this clade are modifications of the palate to accommodate cranial air sinuses, a reduction in the overlap of the dentary and postdentary bones in the mandible to form an intramandibular joint, considerable pneumatization of the vertebral column and the long bones, and a substantial transformation of the hand (a greatly reduced digit V, closely appressed proximal shafts of metacarpals I–III, and deep extensor pits on metacarpals I–III), among other features. Included among the theropods are *Ceratosaurus*, *Allosaurus*, *Velociraptor*, and birds. Tetanurae, Avetheropoda, and Maniraptora are sequential, less inclusive clades within Theropoda (Holtz and Osmólska 2004).

Tibiotarsus (pl. tibiotarsi): A long bone in the leg of a bird, formed by the fusion of the tibia with the proximal bones of the tarsus (the ankle bones).

Titanosauria: A stem-based taxon defined as those sauropods more closely related to *Saltasaurus* than to *Brachiosaurus*, diagnosed by a prominent caudolateral expansion of the caudal end of the sternal plate and an extremely robust radius and ulna. Other titanosaurs include *Argentinosaurus*, *Rapetosaurus*, and *Opisthocoelicaudia* (Upchurch et al. 2004).

Transgression: In a geologic context, the spread of the sea over land areas.

Troodontidae: A stem-based taxon defined as all taxa closer to *Troodon formosus* than to *Velociraptor mongoliensis*. This clade is diagnosed by a neurovascular groove on the outer face of the dentary, a caudal pneumatic foramen in the quadrate, the loss of the basisphenoid recess on the basicranium, a large number of teeth, a sulcus on the dorsal midline of the distal caudal vertebrae, and an asymmetrical ankle and foot, among other features. Other troodontids include *Byronosaurus* and *Saurornithoides* (Makovicky and Norell 2004).

Trophic: Of or relating to nutrition.

Typostrophy: The theory that the pathway of evolution within a taxon goes through stages analogous to the ontogenetic stages of organisms. Thus, there is the origin of a taxon (birth), diversification (growth), degeneration (senescence), and extinction (death).

Tyrannosauroidea: A stem-based taxon defined as the clade of theropods sharing a more recent common ancestor with *Tyrannosaurus rex* than with *Ornithomimus velox*, *Deinonychus antirrhopus*, or *Allosaurus fragilis*. This clade is diagnosed by characteristics of the lower jaw and by pelvic modifications, among other features. Other tyrannosauroids include *Eotyrannus*, *Gorgosaurus*, and *Daspletosaurus* (Holtz 2004).

Vardar Sea: A small oceanic basin, located in present-day Macedonia, that probably opened sometime in the Middle Jurassic and closed in the latest Cretaceous or early Tertiary.

Zero-sum game: A game in which the reward for winning is fixed and can only be won by one of the two players; that is, there can be only one winner and one loser.

REFERENCES

Abel, O. 1912. *Grundzüge der Paläobiologie der Wirbeltiere*. Schweizerbart, Stuttgart. 708 pp.
Abel, O. 1914. Die Tiere der Vorwelt. In: Hertwig, R., and Wettstein, R. von (eds.). *Abstammungslehre Systematik, Palaontologie, Biogeographie*. Teubner, Berlin. Pp. 303–395.
Abel, O. 1929. *Paläobiologie und Stammesgeschichte*, G. Fischer, Jena, Germany. 423 pp.
Adams, R. A. 1998. Evolutionary implications of developmental and functional integration in bat wings. Journal of Zoology 246: 165–174.
Adams, R. A., and Pederson, S. C. 1994. Wings on their fingers. Natural History 103 (1): 49–55.
Alberch, P. 1982. Developmental constraints in evolutionary processes. In: Bonner, J. T. (ed.). Springer-Verlag, Berlin. Pp. 313–332.
Alberch, P., Gould, S. J., Oster, G. F., and Wake, D. B. 1979. Size and shape in ontogeny and phylogeny. Paleobiology 5: 296–317.
Alexander, R. M. 1968. *Animal Mechanics*. University of Washington Press, Seattle. 346 pp.
Alexander, R. M. 1976. Estimates of speeds of dinosaurs. Nature 261: 129–130.
Alexander, R. M. 1985. Mechanics of posture and gait of some large dinosaurs. Zoological Journal of the Linnean Society 83: 1–25.
Alexander, R. M. 1989. *Dynamics of Dinosaurs and Other Extinct Giants*. Columbia University Press, New York. 167 pp.
Alexander, R. M. 1997. Engineering a dinosaur. In: Farlow, J. O., and Brett-Surman, M. K. (eds.). *The Complete Dinosaur*. Indiana University Press, Bloomington. Pp. 415–425.
Allain, R., and Pereda-Suberbiola, X. 2003. Dinosaures de France. Comptes Rendus Palevol 2: 27–44.
Anastasiu, N., and Ciobuca, D. 1989. Non-marine uppermost Cretaceous deposits from Ștei-Densuș region (Hațeg Basin): sketch for a facies model. Revue Roumaine de Géologie, Géographie 33: 43–53.
Anderson, J. F., Hall-Martin, A., and Russell, D. A. 1985. Long bone circum-

ference and weight in mammals, birds, and dinosaurs. Journal of Zoology, Series A, 207: 53–61.

Andrews, C. W. 1913. On some bird remains from the Upper Cretaceous of Transylvania. Geological Magazine 10 (5): 193–196.

Antonescu, E. 1973. Asociaţii palinologice caracteristice unor formaţiuni Cretacice din Munţii Metaliferi. Dări de Seamă al Şedinţelor, 3, Paleontologie 59: 115–169.

Antonescu, E., Lupu, D., and Lupu, M. 1983. Corrélation palynologique du Crétacé terminal du sud-est des Monts Metaliferi et des Dépressions de Haţeg et de Rusca Montană. Anuarul Institutului de Geologie şi Geofizică, Stratigrafie 59: 71–77.

Antunes, M. T., and Sigogneau-Russell, D. 1991. Nouvelles données sur les dinosaures du Crétacé supérieur du Portugal. Comptes Rendus de l'Académie des Sciences de Paris, Série 2, 313: 113–119.

Antunes, M. T., and Sigogneau-Russell, D. 1996. Le Crétacé terminal portugais et son apport au problème de l'extinction des dinosaures. Bulletin du Muséum National d'Histoire Naturelle, Paris, Série 4, 18: 595–606.

Arkhangelsky, M. S., and Averianov, A. O. 2003. [On the find of a primitive hadrosauroid dinosaur (Ornithischia, Hadrosauroidea) in the Cretaceous of the Belgorod Region.] Paleontologicheskii Zhurnal 2003: 60–63. [In Russian.]

Astibia, H., Buffetaut, E., Buscalioni, A. D., Cappetta, H., Corral, C., Estes, R., Garcia-Garmilla, F., Jaeger, J. J., Jimenez-Fuentes, E., Le Loeuff, J., Mazin, J. M., Orue-Etxebarria, X., Pereda-Suberbiola, J., Powell, J. E., Rage, J. C., Rodriguez-Lazaro, J., Sanz, J. L., and Tong, H. 1990. The fossil vertebrates from Laño (Basque Country, Spain); new evidence on the composition and affinities of the Late Cretaceous continental faunas of Europe. Terra Nova 2: 460–466.

Astibia, H., Corral, J. C., Murelaga, X., Orue-Etxebarria, X., and Pereda-Suberbiola, X. (eds.). 1999. *Geology and Paleontology of the Upper Cretaceous Vertebrate-Bearing Beds of the Laño Quarry (Basque-Cantabrian Region, Iberian Peninsula)*. Estudios del Museo de Ciencias Naturales de Alava, Número Especial 1, Vol. 14. 380 pp.

Averianov, A. O., and Yarkov, A. A. 2004. Carnivorous dinosaurs (Saurischia, Theropoda) from the Maastrichtian of the Volgo–Don interfluve, Russia. Paleontological Journal 38: 78–82.

Azzaroli, A. 1982. Insularity and its effects on terrestrial vertebrates: evolutionary and biogeographic aspects. In: Gallitelli, E. M. (ed.). *Palaeontology, Essential of Historical Geology*. S. T. E. M. Mucchi, Modena, Italy. Pp. 194–213.

Bakker, R. T. 1978. Dinosaur feeding behaviour and the origin of flowering plants. Nature 274: 661–663.

Bakker, R. T. 1986. *The Dinosaur Heresies.* William Morrow, New York. 481 pp.

Baraboshkin, E. Yu., Alekseev, A. S., and Kopaevich, L. F. 2003. Cretaceous palaeogeography of the North-Eastern Peri-Tethys. Palaeogeography, Palaeoclimatology, Palaeoecology 196: 177-208.

Barigozzi, C. (ed.). 1982. *Mechanisms of Speciation.* A. R. Liss, New York. 546 pp.

Barnosky, A. D. 2001. Distinguishing the effects of the Red Queen and Court Jester on Miocene mammal evolution in the northern Rocky Mountains. Journal of Vertebrate Paleontology 21: 172-185.

Barron, E. J. 1983. A warm, equable Cretaceous: the nature of the problem. Earth-Science Reviews 19: 305-338.

Barsbold, R. 1982. [Carnivorous dinosaurs from the Cretaceous of Mongolia.] Sovmestnaya Sovetsko-Mongolskaya Paleontologicheskaya Ekspeditsya Trudy 19: 1-117. [In Russian.]

Barsbold, R., and Osmólska, H. 1999. The skull of *Velociraptor* (Theropoda) from the Late Cretaceous of Mongolia. Acta Palaeontologica Polonica 44: 189-219.

Barton, N. H., and Charlesworth, B. 1984. Genetic revolutions, founder effects, and speciation. Annual Review of Ecology and Systematics 15: 133-164.

Basinger, J. 1990. *The It's a Wonderful Life Book.* Knopf, New York. 365 pp.

Bataller, J. R. 1960. Los vertebrados de Cretácico español. Notas y Comunicaciónes del Instituto Geológico y Minero de España 60: 141-164.

Beatty, J. 1995. The evolutionary contingency thesis. In: Wolters, G., and Lennox, J. G. (eds.). *Concepts, Theories, and Rationality in the Biological Sciences.* University of Pittsburgh Press, Pittsburgh. Pp. 45-81.

Bell, G. 1982. *The Masterpiece of Nature: The Evolution and Genetics of Sexuality.* University of California Press, Berkeley. 378 pp.

Bellemain, E., and Ricklefs, R. 2008. Are islands the end of the colonization road? Trends in Ecology and Evolution. 23: 461-468.

Benson, R. B. J. 2010. A description of *Megalosaurus bucklandii* (Dinosauria: Theropoda) from the Bathonian of the UK and the relationships of Middle Jurassic theropods. Zoological Journal of the Linnean Society 158: 882-935.

Benson, R. B. J., Barrett, P. M., Powell, H. P., and Norman, D. B. 2008. The taxonomic status of Megalosaurus bucklandii (Dinosauria, Theropoda) from the Middle Jurassic of Oxfordshire, UK. Palaeontology 51: 419-424.

Benton, M. J. 1990. Phylogeny of the major tetrapod groups: morphological data and divergence dates. Journal of Molecular Evolution 30: 409-424.

Benton, M. J. 2004. Origin and relationships of Dinosauria. In: Weishampel, D. B., Dodson, P., and Osmólska, H. (eds.). *The Dinosauria*, 2nd edition. University of California Press, Berkeley. Pp. 7-20.

Benton, M. J., Cook, E., Grigorescu, D., Popa, E., and Tallódi, E. 1997. Dinosaurs and other tetrapods in an Early Cretaceous bauxite-filled fissure, northwestern Romania. Palaeogeography, Palaeoclimatology, Palaeoecology 130: 275–292.

Benton, M. J., Csiki, Z., Grigorescu, D., Redelstorff, R., Sander, P. M., Stein, K., and Weishampel, D. B. 2010. Dinosaurs and the island rule: The dwarfed dinosaurs from Hațeg Island. Palaeogeography, Palaeoclimatology, Palaeoecology 293: 438–454.

Benton, M. J., and Storrs, G. 1994. Testing the quality of the fossil record: paleontological knowledge is improving. Geology 22: 111–114.

Bergstrom, C. T., and Lachmann, M. 2003. The Red King effect: when the slowest runner wins the coevolutionary race. PNAS: Proceedings of the National Academy of Sciences (USA) 100: 593–598.

Berza, T., Constantinescu, E., and Vlad, Ș.-N. 1998. Upper Cretaceous magmatic series and associated mineralisation in the Carpathian-Balkan orogen. Resource Geology 48: 291–306.

Blackstone, N. W. 1987. Size and time. Systematic Zoology 36: 211–215.

Blueweiss, L., Fox, H., Kudzmal, V., Nakashima, D., Peters, R., and Sams, S. 1978. Relationship between body size and some life history parameters. Oecologia 37: 257–272.

Boccaccio, G. 1350 [reprinted 1999]. *Decameron*. Oxford University Press, Oxford. 752 pp.

Bock, W. J., and Bühler, P. 1996. Nomenclature of Cretaceous birds from Romania. Cretaceous Research 17: 509–514.

Boekschoten, G. J., and Sondaar, P. Y. 1966. The Pleistocene of the Katharo Basin (Crete) and its hippopotamus. Bijdragen tot de Dierkunde 36: 17–44.

Boekschoten, G. J., and Sondaar, P. Y. 1972. On the fossil mammals of Cyprus. Proceedings of the Koninklijke Akademie van Wetenschappen te Amsterdam, Series B, 75: 306–338.

Bojar, A.-V., Grigorecu, D., Ottner, F., and Csiki, Z. 2005. Palaeoenvironmental interpretation of dinosaur- and mammal-bearing continental Maastrichtian deposits, Hațeg Basin, Romania. Geological Quarterly 49: 205–222.

Bonaparte, J. F., and Coria, R. A. 1993. Un nuevo y gigantesco saurópodo titanosaurio de la Formación Río Limay (Albiano–Cenomaniano) de la provincia del Neuquén, Argentina. Ameghiniana 30: 271–282.

Brinkmann, W. 1984. Erster Nachweis eines Hadrosauriers (Ornithischia) aus dem unteren Garumnium (Maastrichtium) des Beckens von Tremp (Provinz Lérida, Spanien). Paläontologisches Zeitschrift 58: 295–305.

Brinkmann, W. 1986. *Rhabdodon* Matheron, 1869 (Reptilia, Ornithischia): proposed conservation by suppression of *Rhabdodon* Fleischmann, 1831 (Reptilia, Serpentes). Bulletin of Zoological Nomenclature 43: 269–272.

Brinkmann, W. 1988. *Zur Fundgeschichte und Systematik der Ornithopoden*

(*Ornithischia, Reptilia*) *aus der Ober-Kreide von Europa*. Documenta Naturae 45. Kanzier, Munich. 156 pp.

Brochu, C. A. 1997. A review of "*Leidyosuchus*" (Crocodyliformes, Eusuchia) from the Cretaceous through Eocene of North America. Journal of Vertebrate Paleontology 17: 679–697.

Brochu, C. A. 1999. Phylogenetics, taxonomy, and historical biogeography of Alligatoroidea. Society of Vertebrate Paleontology Memoir 6: 9–100.

Brodkorb, P. 1978. Catalogue of fossil birds: part 5 (Passeriformes). Bulletin of the Florida State Museum, Biological Sciences 23: 139–228.

Brookes, R. 1763. *The Natural History of Waters, Earths, Stones, Fossils, and Minerals, with their Virtues, Properties, and Medicinal Uses: To Which is added, The Method in which Linnaeus has treated these Subjects*, vol. 5 of *A New and Accurate System of Natural History*. J. Newberry, London. 364 pp. (2nd edition, T. Carnan and F. Newberry, Jr., London, 1772).

Brooks, D. R., and McLennan, D. A. 1991. *Phylogeny, Ecology, and Behavior*. University of Chicago Press, Chicago. 434 pp.

Brown, J. H. 1971. Mammals on mountain tops: nonequilibrium insular biogeography. American Naturalist 105: 467–478.

Brown, J. H. 1975. Geographical ecology of desert rodents. In: Cody, M. L., and Diamond, J. M. (eds.). *The Ecology and Evolution of Communities*. Belknap Press, Cambridge, MA. Pp. 522–536.

Brown, J. H., and Maurer, B. A. 1986. Body size, ecological dominance, and Cope's Rule. Nature 324: 248–250.

Brown, L., and Rockwood, L. L. 1986. On the dilemma of horns. Natural History 95 (7): 54–61.

Buckland, W. 1824. Notice on the *Megalosaurus*, or great fossil lizard of Stonesfield. Transactions of the Geological Society of London 1: 390–396.

Buckling, A., and Rainey, P. B. 2002. Antagonistic coevolution between a bacterium and a bacteriophage. Proceedings of the Royal Society of London B, Biological Sciences 269: 931–936.

Buffetaut, E. 1979. Revision der Crocodylia (Reptilia) aus den Gosau-Schichten (Ober-Kreide) von Österreich. Beiträge zur Paläontologie von Österreich 6: 89–105.

Buffetaut, E. 1987. *A Short History of Vertebrate Palaeontology*. Croom Helm, London. 223 pp.

Buffetaut, E., Cappetta, H., Gayet, M., Martin, M., Moody, R. T. J., Rage, J. C., Taquet, P., and Wellnhofer, P. 1981. Les vertébrés de la partie moyenne du Crétacé en Europe. Cretaceous Research 2: 275–281.

Buffetaut, E., Clottes, P., Cuny, G., Ducrocq, S., Le Loeuff, J., Martin, M., Powell, J. E., Raynaud, C., and Tong H. 1989. Les gisements de dinosaures maastrichtiens de la haute vallée de l'Aude (France): premiers résultats des fouilles de 1989. Comptes Rendus de l'Académie des Sciences de Paris, Série 2, 309: 1723–1727.

Buffetaut, E., Costa, G., Le Loeuff, J., Martin, M., Rage, J.-C., Valentin, X., and Tong H. 1996. An early Campanian vertebrate fauna from the Villeveyrac Basin (Hérault, southern France). Neues Jahrbuch für Geologie und Paläontologie, Monatshefte 1996: 1–16.

Buffetaut, E., Cuny, G., and Le Loeuff, J. 1991. French dinosaurs: the best record in Europe? Modern Geology 16: 17–42.

Buffetaut, E., Cuny, G., and Le Loeuff, J. 1993. The discovery of French dinosaurs. Modern Geology 18: 161–182.

Buffetaut, E., Delfino, M., and Pinna, G. 2001. The crocodilians, pterosaurs, and dinosaurs from the Santonian-Campanian of Villaggio del Pescatore (northeastern Italy): a preliminary report. In: *6th European Workshop on Vertebrate Paleontology, Florence, Italy: Abstract Volume*. Universidà degli Studi di Firenze, Florence, Italy, p. 22.

Buffetaut, E., Grigorescu, D., and Csiki, Z. 2002. A new giant pterosaur with a robust skull from the latest Cretaceous of Romania. Naturwissenschaften 89: 180–184.

Buffetaut, E., Grigorescu, D., and Csiki, Z. 2003. Giant azhdarchid pterosaurs from the terminal Cretaceous of Transylvania (western Romania). *Geological Society of London Special Publications* 217: 91–104.

Buffetaut, E., Laurent, Y., Le Loeuff, J., and Bilotte, M. 1997. A terminal Cretaceous giant pterosaur from the French Pyrenees. Geological Magazine 134: 553–556.

Buffetaut, E., and Le Loeuff, J. 1991a. Late Cretaceous dinosaur faunas of Europe: some correlation problems. Cretaceous Research 12: 159–176.

Buffetaut, E., and Le Loeuff, J. 1991b. Une nouvelle espèce de *Rhabdodon* (Dinosauria, Ornithischia) du Crétacé supérieur de l'Hérault (sud de la France). Comptes Rendus de l'Académie des Sciences de Paris, Série 2, 312: 943–948.

Buffetaut, E., and Le Loeuff, J. 1997. Late Cretaceous dinosaurs from the foothills of the Pyrenees. Geology Today 13: 60–68.

Buffetaut, E., and Le Loeuff, J. 1998. A new giant ground bird from the Upper Cretaceous of southern France. Journal of the Geological Society of London 155: 1–4.

Buffetaut, E., Le Loeuff, J., Mechin, P., and Mechin-Salessy, A. 1995. A large French Cretaceous bird. Nature 377: 110.

Buffetaut, E., Marandat, B., and Sigé, B. 1986. Découverte de dents de deinonychosaures (Saurischia, Theropoda) dans le Crétacé supérieur du sud de la France. Comptes Rendus de l'Académie des Sciences de Paris, Série 2, 303: 1393–1396.

Buffetaut, E., Mechin, P., and Mechin-Salessy, A. 1988. Un dinosaure théropode d'affinités gondwaniennes dans le Crétacé supérieur de Provence. Comptes Rendus de l'Académie des Sciences de Paris, Série 2, 306: 153–158.

Buffetaut, E., Meijer, A. W. F., Taquet, P., and Wouters, G. 1985. New remains of hadrosaurid dinosaurs (Reptilia, Ornithischia) from the Maastrichtian of Dutch and Belgian Limburg. Revue de Paléobiologie 4: 65–70.

Bunzel, E. 1871. Die Reptilfauna der Gosauformation in der Neuen Welt bei Wiener-Neustadt. Abhandlungen der Kaiserlich-Königlischen Geologischen Reichsanstalt Wien 5: 1–18.

Burchfiel, B. C. 1980. Eastern European alpine system and the Carpathian orocline as an example of collision tectonics. Tectonophysics 63: 31–61.

Burness, G. P., Diamond, J., and Flannery, T. 2001. Dinosaurs, dragons, and dwarfs: the evolution of maximal body size. PNAS: Proceedings of the National Academy of Sciences (USA) 98: 14518–14523.

Buscalioni, A. D., Ortega, F., and Vasse, D. 1997. New crocodiles (Eusuchia: Alligatoroidea) from the Upper Cretaceous of southern France. Comptes Rendus de l'Académie des Sciences de Paris, Série 2, 325: 525–530.

Buscalioni, A. D., Ortega, F., Weishampel, D. B., and Jianu, C.-M. 2001. The revision of *Allodaposuchus precedens* from the Upper Cretaceous beds of the Haţeg Basin of Romania. Journal of Vertebrate Paleontology 21: 74–86.

Bush, A. B. G. 1997. Numerical simulation of the Cretaceous Tethys circumglobal current. Science 275: 807–810.

Bush, G. L. 1975. Modes of animal speciation. Annual Review of Ecology and Systematics 6: 334–364.

Butler, R. J., Upchurch, P., and Norman, D. B. 2008. The phylogeny of the ornithischian dinosaurs. Journal of Systematic Palaeontology 6: 1–40.

Cai Z., and Wei F. 1994. On a new pterosaur (*Zhejiangopterus linhaiensis*, gen. et sp. nov.) from Upper Cretaceous in Linhai, Zhejiang, China. Vertebrata PalAsiatica 32: 181–194.

Calder, W. A. 1984. *Size, Function, and Life History*. Harvard University Press, Cambridge, MA. 431 pp.

Carlquist, S. 1974. *Island Biology*. Columbia University Press, New York. 660 pp.

Carpenter, K. 1997. Ankylosaurs. In: Farlow, J. O., and Brett-Surman, M. K. (eds.). *The Complete Dinosaur*. Indiana University Press, Bloomington, IN. Pp. 307–316.

Carrier, M. 1995. Evolutionary change and lawlikeness. In: Wolters, G., and Lennox, J. G. (eds.). *Concepts, Theories, and Rationality in the Biological Sciences*. University of Pittsburgh Press, Pittsburgh. Pp. 83–97.

Casanovas, M. L. 1992. Novedades en el registro fósil de dinosaurios de Levante español. Zubia 10: 139–151.

Casanovas, M. L., Pereda-Suberbiola, X., Santafé, J. V., and Weishampel, D. B. 1999a. First lambeosaurine hadrosaurid from Europe. Geological Magazine 136: 205–211.

Casanovas, M. L., Pereda-Suberbiola, X., Santafé, J. V., and Weishampel, D. B. 1999b. A primitive euhadrosaurian dinosaur from the uppermost Cretaceous of the Ager Syncline (southern Pyrenees, Catalonia). Geologie en Mijnbouw 78: 345–356.

Casanovas, M. L., Santafé, J. L., Sanz, J. L., and Buscalioni, A. D. 1987. Arcosaurios (Crocodilia, Dinosauria) del Cretácico superior de la Conca de Tremp (Lleida, España). Estudios Geológicos, Vol. Extraordinario Galve-Tremp: 95–110.

Casanovas-Cladellas, M. L., and Santafé-Llopis, J. V. 1993. Presencia de titanosauridos (Dinosauria) en el Cretácico superior de Fontllonga (Lleida, España). Treballs del Museu de Geología de Barcelona 3: 67–80.

Casanovas-Cladellas, M. L., Santefé-Llopis, J. V., and Isidro-Llorens, A. 1993. *Pararhabdodon isonense* n. gen. n. sp. (Dinosauria): estudio morfológico, radio-tomografico, y consideraciónes biomecánicas. Paleontologia i Evolució 26–27: 121–131.

Casanovas-Cladellas, M. L., Santafé-Llopis, J. V., Sanz, J. L., and Buscalioni, A. 1985. *Orthomerus* (Hadrosauridae, Ornithopoda) del Cretácico superior del yacimiento de (Tremp, España). Palaeontologia i Evolució 19: 155–162.

Casanovas-Cladellas, M. L., Santafé-Llopis, J. V., and Sanz-Garcia, J. L. 1988. La primera resta fòssil d'un Teròpode (Saurischia, Dinosauria) en el Cretaci superior de la Conca de Tremp (Lleida, Espanya). Paleontologia i Evolució 22: 7–81.

Case, T. J. 1978. A general explanation for insular body size trends in terrestrial vertebrates. Ecology 59: 1–18.

Castanet, J. 1987. La squelettochronologie chez les Reptiles, III: applications. Annales des Sciences Naturelle, Zoologie, Série 13 (8): 157–172.

Castanet, J., Francillon-Vieillot, H., Meunier, F. J., and Ricqlès, A. de. 1993. Bone and individual aging. In: Hall, B. K. (ed.). *Bone*, vol. 7 of *Bone Growth*. CRC Press, London. Pp. 245–283.

Cheylan, G. (ed.). 1994. *Dinosaures en Provence*. Éditions Edisud, Aix-en-Provence, France. 71 pp.

Chiappe, L. M. 1995. The first 85 million years of avian evolution. Nature 378: 349–355.

Chiappe, L. M., Coria, R. A., Dingus, L., Jackson, F., Chinsamy, A., and Fox, M.

1998. Sauropod dinosaur embryos from the Late Cretaceous of Patagonia. Nature 396: 258–261.

Chiappe, L. M., and Dingus, L. 2001. *Walking on Eggs*. Scribner, New York. 224 pp.

Chiappe, L. M., Norell, M. A., and Clark, J. M. 2002. The Cretaceous, short-armed Alvarezsauridae: *Mononykus* and its kin. In: Chiappe, L. M., and Witmer, L. M. (eds.). *Mesozoic Birds: Above the Heads of Dinosaurs*. University of California Press, Berkeley. Pp. 87–120.

Chiappe, L. M., Salgado, L., and Coria, R. A. 2001. Embryonic skulls of titanosaur sauropod dinosaurs. Science 293: 2444–2446.

Chiappe, L. M., and Witmer, L. M. (eds.). 2002. *Mesozoic Birds: Above the Heads of Dinosaurs*. University of California Press, Berkeley. 576 pp.

Chinsamy, A. 1995. Ontogenetic changes in the bone histology of the Late Jurassic ornithopod *Dryosaurus lettowvorbecki*. Journal of Vertebrate Paleontology 15: 96–104.

Chinsamy, A., and Hillenius, W. J. 2004. Physiology of nonavian dinosaurs. In: Weishampel, D. B., Dodson, P., and Osmólska, H. (eds.). *The Dinosauria*, 2nd edition. University of California Press, Berkeley. Pp. 643–659.

Chinsamy-Turan, A. 2005. *The Microstructure of Dinosaur Bone*. Johns Hopkins University Press, Baltimore. 195 pp.

Christie, I. (ed.). 1999. *Gilliam on Gilliam*. Faber and Faber, New York. 208 pp.

Clark, J. M., Maryañska, T., and Barsbold, R. Therizinosauroidea. 2004. In: Weishampel, D. B., Dodson, P., and Osmólska, H. (eds.). *The Dinosauria*, 2nd edition. University of California Press, Berkeley. Pp. 151–164.

Clark, J. M., Norell, M. A., and Chiappe, L. M. 1999. An oviraptorid skeleton from the Late Cretaceous of Ukhaa Tolgod, Mongolia, preserved in an avianlike brooding position over an oviraptorid nest. American Museum Novitates 3265: 1–36.

Cloutier, R. 1991. Patterns, trends, and rates of evolution within the Actinistia. Environmental Biology of Fishes 32: 23–58.

Codrea, V., and Dica, P. 2005. Upper Cretaceous-lowermost Miocene lithostratigraphic units exposed in Alba Iulia-Sebes-Vinţu de Jos area (SW Transylvanian Basin). Studia Universitatis Babeş-Bolyai, Geologia 50: 19–26.

Codrea, V., and Godefroit, P. 2008. New Late Cretaceous dinosaur findings from northwestern Transylvania (Romania). Comptes Rendus Palevol 7: 289–295.

Codrea, V., Hosu, A., Filipescu, S., Vremir, M., Dica, P., Săsăran, E., and Tanţău, I. 2001. Aspecte ale sedimentaţiei Cretacic superioare din Aria Alba Iulia-Sebeş (Jud. Alba). Studii şi Cercetări de Geologie-Geografie 6: 63–68.

Codrea, V., Săsăran, E., and Dica, P. 2002. Vurpăr (Vinţu de Jos, Alba district).

In: *7th European Workshop of Vertebrate Palaeontology, Sibiu, Romania: Abstracts Volume and Excursions Field Guide*. Pp. 60–62.

Codrea, V., Smith, T., Dica, P., Folie, A., Garcia, G., Godefroit, P., and Van Itterbeeck, J. 2002. Dinosaur egg nests, mammals, and other vertebrates from a new Maastrichtian site of the Haţeg Basin (Romania). Comptes Rendus Palevol 1: 173–180.

Codrea, V., Vremir, M., Jipa, C., Godefroit, P., Csiki, Z., Smith, T., and Fărcaş, C. 2010. More than just Nopcsa's Transylvanian dinosaurs: a look outside the Haţeg Basin. Palaeogeography, Palaeoclimatology, Palaeoecology 293: 391–405.

Colbert, E. H. 1962. The weights of dinosaurs. American Museum Novitates 2076: 1–16.

Colbert, E. H. 1968. *Men and Dinosaurs: The Search in Field and Laboratory*. Dutton, New York. 283 pp.

Company, J., Galobart, A., and Gaete, R. 1998. First data on the hadrosaurid dinosaurs (Ornithischia, Dinosauria) from the Upper Cretaceous of Valencia, Spain. Oryctos 1: 121–126.

Company, J., Pereda-Suberbiola, X., Ruiz-Omenaca, J. I., and Buscalioni, A. D. 2005. A new species of *Doratodon* (Crocodyliformes, Ziphosuchia) from the Late Cretaceous of Spain. Journal of Vertebrate Paleontology 25: 343–353.

Conrad, J. L. 2008. Phylogeny and systematics of Squamata (Reptilia) based on morphology. Bulletin of the American Museum of Natural History 310: 1–182.

Conrad von Hötzendorff, F. 1921–1925. *Aus Meiner Dienstzeit, 1906-1918*. 5 vols. Rikola Verlag, Vienna.

Coombs, W. P., and Maryañska, T. 1990. Ankylosauria. In: Weishampel, D. B., Dodson, P., and Osmólska, H. (eds.). *The Dinosauria*. University of California Press, Berkeley. Pp. 456–482.

Cope, E. D. 1896. *The Primary Factors of Organic Evolution*. Open Court, Chicago. 547 pp.

Coria, R. A., and Salgado, L. 1995. A new giant carnivorous dinosaur from the Cretaceous of Patagonia. Nature 377: 224–226.

Corti, E. C. 1936. *Elizabeth, Empress of Austria* (translated by Phillips, C. A.). Thornton Butterworth, London. 416 pp.

Costea, I., Comşa, D., and Vinogradov, C. 1978. Microfaciesurile Cretacicului inferior din platforma Moesică. Studii şi Cercetări de Geologie, Geofizică, Geografie, Serie Geologie 23: 299–311.

Courtenay-Latimer, M. 1979. My story of the first coelacanth. Occasional Papers of the California Academy of Science 134: 6–10.

Cousin, R., Breton, G., Fournier, R., and Watté, J.-P. 1994. Dinosaur egg laying and nesting in France. In: Carpenter, K., Hirsch, K. F., and Horner, J. R. (eds.). *Dinosaur Eggs and Babies*. Cambridge University Press, New York. Pp. 56–74.

Coyne, J. A. 1992. Genetics and speciation. Nature 355: 511–515.

Csiki, Z. 1997. Legături paleobiogeographice ale faunei de vertebrate continentale Maastrichtian superioare din Bazinul Hațeg. Nymphaea 23–25: 45–68.

Csiki, Z. 1999. New evidence of armoured titanosaurids in the Late Cretaceous—*Magyarosaurus dacus* from the Hațeg Basin (Romania). Oryctos 2: 93–99.

Csiki, Z. 2008. Inset borings in dinosaur bones from the Maastrichtian of the Hațeg Basin, Romania—paleoecological and paleoclimatic implications. In: Csiki, Z. (ed.). *Mesozoic and Cenozoic Vertebrates and Paleoenvironments*. Ars Docendi, Bucharest. Pp. 95–104.

Csiki, Z., Codrea, V., Jipa-Murzea, C., and Godefroit, P. 2010. A partial titanosaur (Sauropoda, Dinosauria) skeleton from the Maastrichtian of Nălaț-Vad, Hațeg Basin, Romania. Neues Jahrbuch für Geologie und Paläontologie, Abhandlungen. doi:10.112/0077-7749/2010/0098.

Csiki, Z., and Grigorescu, D. 1998. Small theropods from the Late Cretaceous of the Hațeg Basin (western Romania)—an unexpected diversity at the top of the food chain. Oryctos 1: 87–104.

Csiki, Z., and Grigorescu, D. 2000. Teeth of multituberculate mammals from the Late Cretaceous of Romania. Acta Palaeontologica Polonica 45: 85–90.

Csiki, Z., and Grigorescu, D. 2001. Paleobiogeographic implications of the fossil mammals from the Maastrichtian of the Hațeg Basin, Romania. Acta Palaeontologica Romaniae 3: 89–95.

Csiki, Z., and Grigorescu, D. 2006a. A hátszegi "*Magyarosaurus*" *hungaricus* Huene (Dinosauria: Sauropoda) reviziója. *8th Mining, Metallurgy, and Geology Conference: Extended Abstracts*. Erdélyi Magyar Múszaki Todományos Társaság, Cluj-Napoca, Romania. Pp. 65–70.

Csiki, Z., and Grigorescu, D. 2006b. Maastrichtian multituberculates of the Hațeg Basin, Romania—implications for multituberculate phylogeny, evolution, and paleobiogeography. In Barrett, P. M., and Evans, S. E. (eds.). *Ninth International Symposium on Mesozoic Terrestrial Ecosystems and Biota: Abstracts and Proceedings*. Natural History Museum, London. Pp. 29–32.

Csiki, Z., and Grigorescu, D. 2007. The "Dinosaur Island"—new interpretation of the Hațeg Basin vertebrate faauna after 110 years. Sargetia 20: 5–26.

Csiki, Z., Grigorescu, D., and Rücklin, M. 2005. A new multituberculate specimen from the Maastrichtian of Pui, Romania, and reassessment of affinities of *Barbatodon*. Acta Palaeontologica Romaniae 5: 73–86.

Csiki, Z., Grigorescu, D., and Weishampel. D. B. 2007. A new titanosaur sauropod (Dinosauria: Saurischia) from Upper Cretaceous of the Hațeg Basin

(Romania). In: Le Loeuff, J. (ed.). *5th Meeting of the European Association of Vertebrate Palaeontologists, Carcassone-Espéraza, France: Abstracts Volume.* Musée des Dinosaures, Espéraza, France. P. 17.

Csiki, Z., Vremir, M., Brusatte, S. L., and Norell, M. A. 2010. An aberrant island-dwelling theropod dinosaur from the Late Cretaceous of Romania. PNAS: Proceedings of the National Academy of Science (USA) 107: 15357–15361. doi/10.1073/pnas1006970107.

Csontos, L., and Vörös, A. 2005. Mesozoic plate tectonic reconstruction of the Carpathian region. Palaeogeography, Palaeoclimatology, Palaeoecology 210: 1–56.

Curry, K. A. 1999. Ontogenetic histology of *Apatosaurus* (Dinosauria: Sauropoda): new insights on growth rates and longevity. Journal of Vertebrate Paleontology 19: 654–665.

Curry, K. A. 2001. The evolutionary history of the Titanosauria. PhD diss., State University of New York, Stony Brook. 552 pp.

Curry Rogers, K. A. 2005. The evolutionary history of the Titanosauria. In: Curry Rogers, K. A., and Wilson, J. A. (eds.). *The Sauropods: Evolution and Paleobiology.* University of California Press, Berkeley. Pp. 50–103.

Curry Rogers, K. A., and Forster, C. A. 2001. The last of the dinosaur titans: a new sauropod from Madagascar. Nature 412: 530–524.

Dalla Vecchia, F. M. 1997. Terrestrial tetrapod evidence on the Norian (Late Triassic) and Cretaceous carbonate platforms of northern Adriatic region (Italy, Slovenia, and Croatia). Sargetia 17: 177–201.

Dalla Vecchia, F. M. 2001. Terrestrial ecosystems on the Mesozoic peri-Adriatic carbonate platforms: the vertebrate evidence. Publicación Especial, Asociación Paleontologica Argentina 7: 77–83.

Dalla Vecchia, F. M. 2006. *Telmatosaurus* and the other hadrosaurids of the Cretaceous European archipelago: an overview. Natura Nascosta 32: 1–55.

Dalla Vecchia, F. M. 2009. *Tethyshadros insularis*, a new hadrosauroid dinosaur (Ornithischia) from the Upper Cretaceous of Italy. Journal of Vertebrate Paleontology 29: 1100–1116.

Dal Sasso, C. 2003. Dinosaurs of Italy. Comptes Rendus Palevol 2: 45–66.

Damuth, J. 1993. Cope's Rule, the Island Rule, and the scaling of mammalian population density. Nature 365: 748–750.

Danilov, I. G., and Parham, J. F. 2008. A reassessment of some poorly known turtles from the Middle Jurassic of China, with comments on the antiquity of extant turtles. Journal of Vertebrate Paleontology 28: 306–318.

Darwin, C. R. 1859. *On the Origin of Species.* John Murray, London. 502 pp.

Dean, D. R. 1999. *Gideon Mantell and the Discovery of Dinosaurs.* Cambridge University Press, Cambridge. 290 pp.

Debeljak, I., Košir, A., and Otoničar, B. 1999. A preliminary note on dinosaurs and non-dinosaurian reptiles from the Upper Cretaceous carbonate platform succession at Kozina (SW Slovenia): razprave 4. Razreda Sazu 40: 3–25.

Delair, J. B., and Sarjeant, W. A. S. 1975. The earliest discoveries of dinosaurs. Isis 66: 5–25.

Delamare-Deboutteville, C., and Botosancanu, I. 1970. *Formes Primitives Vivantes*. Harmann, Paris. 232 pp.

Delfino, M., Codrea, V., Folie, A., Dica, P., Godefroit, P., and Smith, T. 2008. A complete skull of *Allodaposuchus precedens* Nopcsa, 1928 (Eusuchia) and a reassessment of the morphology of the taxon based on the Romanian remains. Journal of Vertebrate Paleontology 28: 111–122.

Delfino, M., Martin, J. E., and Buffetaut, E. 2008. A new species of Acynodon (Crocodylia) from the Upper Cretaceous (Santonian–Campanian) of Villaggio del Pescatore, Italy. Palaeontology 51: 1091–1106.

Delfino, M., Piras, P., and Smith, T. 2005. Anatomy and phylogeny of the gavialoid crocodylian *Eosuchus lerichei* from the Paleocene of Europe. Acta Palaeontologica Polonica 50: 565–580.

Demment, M. W., and Van Soest, P. J. 1985. A nutritional explanation for body-size patterns of ruminant and nonruminant herbivores. American Naturalist 125: 641–672.

de Queiroz, K., and Gauthier, J. 1990. Phylogeny as a central principle in taxonomy: phylogenetic definitions of taxon names. Systematic Zoology 39: 307–322.

Dercourt, J., Ricou, L.-E., and Vrielynck, B. (eds.). 1993. *Atlas Tethys Palaeoenvironmental Maps*. Gauthier-Villars, Paris. 307 pp.

Dercourt, J., Zonenshain, L. P., Ricou, L.-E., Kazmin, V. G., Le Pichon, X., Knipper, A. L., Grandjacquet, C., Sbortshikov, L. M., Geyssant, J., Lepvrier, C., Pechersky, D. H., Boulin, J., Sibuet, J.-C., Savostin, L. A., Sorokhtin, O., Westphal, M., Bazhenov, M. L., Lauer, J. P., and Biju-Duval, B. 1986. Geological evolution of the Tethys belt from the Atlantic to the Pamirs since the Lias. Tectonophysics 123: 241–315.

Desmond, A. 1979. Designing the dinosaur: Richard Owen's response to Robert Edward Grant. Isis 70: 224–234.

Desmond, A. 1982. *Archetypes and Ancestors: Palaeontology in Victorian London 1850–1875*. University of Chicago Press, Chicago. 287 pp.

Desmond, A. 1989. *The Politics of Evolution*. University of Chicago Press, Chicago. 503 pp.

Dietl, G. R. 2003. Coevolution of a marine gastropod predator and its dangerous bivalve prey. Biological Journal of the Linnean Society 80: 409–436.

Dietz, R. S., and Holden, J. C. 1970. Reconstruction of Pangea: breakup and dispersion of continents, Permian to present. Journal of Geophysical Research 75: 4939–4956.

Dincă, A., Tocorjescu, M., and Stilla, A. 1971. Despre vîrsta depozitelor continentale cu dinozaurieni din Bazinele Hațeg și Rusca Montană. Dări de Seamă al Ședințelor, 4, Stratigrafie 58: 83–93.

Dodson, P. 1990. Sauropod paleoecology. In: Weishampel, D. B., Dodson, P., and Osmólska, H. (eds.). *The Dinosauria*. University of California Press, Berkeley. Pp. 402–407.

Dodson, P., Krause, D. W., Forster, C. A., Sampson, S. D., and Ravoavy, R. 1998. Titanosaurid (Sauropoda) osteoderms from the Late Cretaceous of Madagascar. Journal of Vertebrate Paleontology 18: 563–568.

Dușa, A. 1974. Aspecte ale formării cărbunilor din Bazinul Rusca Montană. Studia Universitatis Babeș-Bolyai, Geologia-Mineralogia 19: 36–43.

Edinger, T. 1955. Personalities in paleontology: Nopcsa. Society of Vertebrate Paleontology News Bulletin 43: 35–39.

Edmonds, J. M. 1979. The founding of the Oxford Readership in Geology, 1818. Notes and Records of the Royal Society of London 30: 141–167.

Ehrenberg, K. 1975. *Othenio Abel's Lebensweg*. Österreichische Hochshülerschaft, Universität Wien, Vienna. 162 pp.

Eldredge, N., and Stanley, S. M. (eds.). 1984. *Living Fossils*. Springer-Verlag, New York. 291 pp.

Elsie, R. 1999. The Viennese scholar who almost became King of Albania: Baron Franz Nopcsa and his contribution to Albanian studies. East European Quarterly 23: 327–345.

Elzanowski, A. 1983. Birds in Cretaceous ecosystems. Acta Paleontologica Polonica 28: 75–92.

Emerson, S. B. 1986. Heterochrony and frogs: the relationship of a life history trait to morphological form. American Naturalist 127: 167–183.

Endler, J. A. 1977. *Geographic Variation, Speciation, and Clines*. Princeton University Press, Princeton, NJ. 246 pp.

Erdmann, M. V., Caldwell, R. L., and Kasim Moosa, M. K. 1998. Indonesian "king of the sea" discovered. Nature 395: 335.

Erickson, G. M. 2005. Assessing dinosaur growth patterns: a microscopic revolution. Trends in Ecology and Evolution 20: 677–684.

Erickson, G. M., Currie, P. J., Inouye, B. D., and Winn, A. A. 2006. Tyrannosaur life tables: the first look at non-avian dinosaur population biology. Science 313: 213–217.

Erickson, G. M., Makovicky, P. J., Currie, P. J., Norell, M. A., Yerby, S. A., and Brochu, C. A. 2004. Gigantism and comparative life-history parameters of tyrannosaurid dinosaurs. Nature 430: 772–775.

Erickson, G. M., Makovicky, P. J., Inouye, B. D., Zhou C.-F., and Zhou C.-Q. 2009. A life table for Psittacosaurus lujiatunensis: initial insights into ornithischian dinosaur population biology. Anatomical Record 292: 1514–1521.

Erickson, G. M., and Tumanova, T. A. 2000. Growth curve of *Psittacosaurus mongoliensis* Osborn (Ceratopsia: Psittacosauridae) inferred from long bone histology. Zoological Journal of the Linnean Society 130: 551–566.

Farlow, J. O. 1981. Estimates of dinosaur speeds from a new trackway site in Texas. Nature 294: 747–748.

Farris, J. S. 1970. Methods for computing Wagner trees. Systematic Zoology 19: 83–92.

Fastovsky, D. E., and Weishampel, D. B. 2009. *Dinosaurs: A Concise Natural History*. Cambridge University Press, New York. 379 pp.

Fejfar, O., Košťák, M., Kvaček, J., Mazuch, M., and Moučka, M. 2005. First Cenomanian dinosaur from Central Europe (Czech Republic). Acta Palaeontologica Polonica 50: 295–300.

Folie, A., and Codrea, V. 2005. New lissamphibians and squamates from the Maastrichtian of the Haţeg Basin, Romania. Acta Palaeontologica Polonica 50: 57–71.

Forey, P. 1988. Golden jubilee for the coelacanth *Latimeria chalumnae*. Nature 336: 727–732.

Forey, P. 1998. A home away from home for coelacanths. Nature 395: 319–320.

Foster, J. B. 1964. The evolution of mammals on islands. Nature 202: 234–235.

Fritsch, A. 1881. Über die Endeckung von Vögelresten in der Böhmischen Kreideformation. Sitzungsberichte der Königlich-Böhmischen Gesellschaft der Wissenschaften 1880: 275–276.

Fritsch, A., and Bayer, F. 1905. *Neue Fische und Reptilien aus der Böhmischen Kreideformation*. Selbstverlag, Prague. 34 pp.

Frost, D. R., Grant, T., Faivovich, J., Bain, R. H., Haas, A., Haddad, C. F. B., Sá, R. O. de, Channing, A., Wilkinson, M., Donnellan, S. C., Raxworthy, C. J., Campbell, J. A., Blotto, B. L., Moler, P., Drewes, R. C., Nussbaum, R. A., Lynch, J. D., Green, D. M., and Wheeler, W. C. 2006. The amphibian tree of life. Bulletin of the American Museum of Natural History 297: 1–291.

Futuyma, D. J. 1998. *Evolutionary Biology*. Sinauer Associates, Sunderland, MA. 763 pp.

Gaffney, E. S., and Meylan, P. A. 1988. A phylogeny of turtles. In: Benton, M. J. (ed.). *Amphibians, Reptiles, Birds*, vol. 1 of *The Phylogeny and Classification of the Tetrapods*. Systematic Association Special Vol. 35. Clarendon Press, Oxford. Pp. 157–219.

Gaffney, E. S., and Meylan, P. A. 1992. The Transylvanian turtle, *Kallokibotion*, a primitive cryptodire of Cretaceous age. *American Museum Novitates* 3040: 1–37.

Galton, P. M. 1996. Notes on Dinosauria from the Upper Cretaceous of Portugal. *Neues Jahrbuch für Geologie und Paläontologie Monatshefte* 1996: 83–90.

Galton, P. M., and Upchurch, P. 2004. Prosauropoda. In: Weishampel, D. B., Dodson, P., and Osmólska, H. (eds.). *The Dinosauria*, 2nd edition. University of California Press, Berkeley. Pp. 232–258.

Garcia, G., Pincemaille, M., Vianey-Liaud, M., Marandat, B., Lorenz, E., Cheylan, G., Cappetta, H., Michaux, J., and Sudre, J. 1999. Découverte du premier squelette presque complet de *Rhabdodon priscus* (Dinosauria, Ornithopoda) du Maastrichtien supérieur de Provence. *Comptes Rendus de l'Académie des Sciences Paris*, Série 2, 328: 415–421.

Gardner, J. D. 2000. Albanerpetontid amphibians from the Upper Cretaceous (Campanian and Maastrichtian) of North America. *Geodiversitas* 22: 349–388.

Gardner, J. D. 2002. Monophyly and intra-generic relationships of *Albanerpeton* (Lissamphibia; Albanerpetontidae). *Journal of Vertebrate Paleontology* 22: 12–22.

Gardner, J. D., Evans, S. E., and Sigogneau-Russell, D. 2003. New albanerpetontid amphibians from the Early Cretaceous of Morocco and Middle Jurassic of England. *Acta Palaeontologica Polonica* 48: 301–319.

Gauthier, J. A., Kluge, A. G., and Rowe, T. 1988. Amniote phylogeny and the importance of fossils. *Cladistics* 4: 105–209.

Gheerbrant, E., Codrea, V., Hosu, A., Sen, S., Guernet, C., Lapparent de Broin, F. de, and Riveline, J. 2000. Découverte en Transylvanie (Roumanie) de gisements à vertébrés dans les calcaires de Rona (Thanétien ou Sparnacien): les plus anciens mammifères cénozoïques d'Europe orientale. *Eclogae Geologicae Helvetiae* 92: 517–535.

Giddings, L. V., Kaneshiro, K. Y., and Anderson, W. W. (eds.). 1997. *Genetics, Speciation, and the Founder Principle*. Oxford University Press, Oxford. 400 pp.

Gillette, D. D. 1994. *Seismosaurus: The Earth Shaker*. Columbia University Press, New York. 205 pp.

Ginsburg, L. 1984. Nouvelles lumières sur les ossements fossiles autrefois attribués au géant Theutobochus. *Annales de Paléontologie* 70: 181–219.

Glut, D. E., and Brett-Surman, M. K. 1997. Dinosaurs and the media. In: Farlow, J. O., and Brett-Surman, M. K. (eds.). *The Complete Dinosaur*. Indiana University Press, Bloomington. Pp. 675–697.

Godefroit, P., Bolotsky, Y., and Alifanov, V. 2003. A remarkable hollow-crested hadrosaur from Russia: an Asian origin for lambeosaurines. Comptes Rendus Palevol 2: 143–151.

Godefroit, P., Codrea, V., and Weishampel, D. B. 2009. Osteology of *Zalmoxes shqiperorum* (Dinosauria, Ornithopoda), based on new specimens from the Upper Cretaceous of Nălaţ-Vad (Romania). Geodiversitas 31: 525–553.

Godefroit, P., Dong Z.-M., Bultynck, P., and Feng L. 1998. Sino-Belgium Cooperation Program, Cretaceous dinosaurs and mammals from Inner Mongolia, I: new *Bactrosaurus* (Dinosauria: Hadrosauroidea) material from Iren Dabasu (Inner Mongolia, P. R. China). Bulletin de l'Institut Royal des Sciences Naturelles de Belgique 68 (Suppl.): 1–70.

Godefroit, P., and Motchurova-Dekova, N. 2010. Latest Cretaceous hadrosauroid (Dinosauria: Ornithopoda) remains from Bulgaria. Comptes Rendus Palevol 9: 163–169.

Godfrey, L. R., and Sutherland, M. R. 1996. Paradox of peramorphic paedomorphosis: heterochrony and human evolution. American Journal of Physical Anthropology 99: 17–42.

Goldman, W. 2000. *The Princess Bride*. Ballantine, New York. 399 pp.

Gould, S. J. 1966. Allometry and size in ontogeny and phylogeny. Biological Reviews 41: 587–640.

Gould, S. J. 1974. The evolutionary significance of "bizarre" structures: antler size and skull size in the "Irish Elk," *Megaloceras gigantans*. Evolution 28: 191–220.

Gould, S. J. 1975. On the scaling of tooth size in mammals. American Zoologist 15: 351–362.

Gould, S. J. 1977. *Ontogeny and Phylogeny*. Belknap Press, Cambridge, MA. 501 pp.

Gould, S. J. 1989. *Wonderful Life*. W. W. Norton, New York. 347 pp.

Gould, S. J., and Vrba, E. S. 1982. Exaptation—a missing term in the science of form. Paleobiology 8: 4–15.

Grant, P. R. 1965. The adaptive significance of some size trends in island birds. Evolution 19: 355–367.

Greene, M. 1983. *Geology in the Nineteenth Century: Changing View of a Changing World*. Cornell University Press, Ithaca, NY. 324 pp.

Grigorescu, D. 1982. Cadrul stratigrafic şi paleoecologic al depozitelor continentale cu dinozauri din Bazinul Haţeg. Sargetia 13: 37–47.

Grigorescu, D. 1983. A stratigraphic, taphonomic, and palaeoecologic approach to a "forgotten land": the dinosaur-bearing deposits from the Haţeg Basin (Transylvania—Romania). Acta Palaeontologica Polonica 28: 103–121.

Grigorescu, D. 1984. New tetrapod groups in the Maastrichtian of the Haţeg

Basin: coelurosaurs and multituberculates. In: Reif, W.-E., and Westphal, F. (eds.). *Third Symposium on Mesozoic Terrestrial Ecosystems, Tübingen. 1984: Short Papers.* Attempto Verlag, Tübingen, Germany. Pp. 99–104.

Grigorescu, D. 1992a. Nonmarine Cretaceous formations of Romania. In: Mateer, N. J., and Chen P.-J. (eds.). *Aspects of Nonmarine Cretaceous Geology.* China Ocean Press, Beijing. Pp. 142–164.

Grigorescu, D. 1992b. The latest Cretaceous dinosaur eggs and embryos from the Haţeg Basin, Romania. Revue de Paléobiologie, Vol. Spécial 7: 95–99.

Grigorescu, D., and Csiki, Z. 2006. Ontogenetic development of *Telmatosaurus transsylvanicus* (Ornithischia: Hadrosauria) from the Maastrichtian of the Haţeg Basin (Romania)—evidence from the limb bones. Hantkeniana 5: 20–26.

Grigorescu, D., Garcia, G., Csiki, Z., Codrea, V., and Bojar, A.-V. 2010. Uppermost Cretaceous megaloolithid eggs from the Haţeg Basin, Romania, associated with hadrosaur hatchlings: search for explanation. Palaeogeography, Palaeoclimatology, Palaeoecology 293: 360–374.

Grigorescu, D., and Hahn, G. 1987. The first multituberculate teeth from the Upper Cretaceous of Europe (Romania). Geologica et Palaeontologica 21: 237–243.

Grigorescu, D., Hartenberger, J.-L., Rădulescu, C., Samson, P., and Sudre, J. 1985. Découverte de mammifères et dinosaures dans le Crétacé supérieur de Pui (Roumanie). Comptes Rendus de l'Académie des Sciences de Paris, Série 2, 301: 1365–1368.

Grigorescu, D., and Kessler, E. 1980. A new specimen of *Elopteryx nopcsai* Andrews from the dinosaurian beds of Haţeg Basin. Revue Roumaine de Géologie, Géophysique, et Géographie, Série Géologie 24: 171–175.

Grigorescu, D., Şeclamen, M., Norman, D. B., and Weishampel, D. B. 1990. Dinosaur eggs from Romania. Nature 346: 417.

Grigorescu, D., Venczel, M., Csiki, Z., and Limberea, R. 1999. New latest Cretaceous microvertebrate fossil assemblages from the Haţeg Basin (Romania). Geologie en Mijnbouw 78: 301–314.

Grigorescu, D., Weishampel, D., Norman, D., Şeclamen, M., Rusu, C., Baltreş, A., and Teodorescu, V. 1994. The hadrosaurid eggs from the Haţeg Basin (Late Cretaceous, Romania). In: Carpenter, K., Hirsch, K., and Horner, J. R. (eds.). *Dinosaur Eggs and Babies.* Cambridge University Press, New York. Pp. 75–87.

Groza, I. 1982. Rezultatele preliminare ale cercetărilor întreprinse de către Muzeul Judeţean Hunedoara-Deva în stratele cu dinosauri de la Sînpetru-Haţeg. Sargetia 13: 49–66.

Haeckel, E. 1866. *Generelle Morphologie der Organismen: Allgemeine Grundzüge der Organischen Formen-Wissenschaft, Mechanisch Begründet Durch die von Charles Darwin Reformierte Descendenz-Theorie*. 2 vols. Georg Reimer, Berlin. 574 pp., 462 pp.

Hairston, N. G. 1951. Interspecies competition and its probable influence upon the vertical distribution of Appalachian salamanders of the genus *Plethodon*. Ecology 32: 266–274.

Haldane, J. B. S. 1956. The relation between density regulation and natural selection. Proceedings of the Royal Society of London B, Biological Sciences 145: 306–308.

Haldane, J. B. S. 1957. The cost of natural selection. Journal of Genetics 55: 511–524.

Hallam, A. 1992. *Phanerozoic Sea-Level Changes*. Columbia University Press. New York. 266 pp.

Halstead, L. B. 1970. *Scrotum humanum* Brookes 1763: the first named dinosaur. Journal of Insignificant Research 5: 14–15.

Halstead, L. B., and Sarjeant, W. A. S. 1993. *Scrotum humanum* Brookes: the earliest name for a dinosaur? Modern Geology 18: 221–224.

Hamann, B. 1986. *The Reluctant Empress*. Knopf, New York. 410 pp.

Hanken, J., and Wake, D. B. 1993. Miniaturization of body size: organismal consequences and evolutionary significance. Annual Review of Ecology and Systematics 24: 501–519.

Haq, B. U., Hardenbol, J., and Vale, P. R. 1988. Mesozoic and Cenozoic chronostratigraphy and cycles of sea-level change. Society of Economic Paleontologists and Mineralogists, Special Publication 42: 71–108.

Harrison, C. J. O., and Walker, C. A. 1975. The Bradycnemidae, a new family of owls from the Upper Cretaceous of Romania. Palaeontology 18: 563–570.

Head, J. J. 1998. A new species of basal hadrosaurid (Dinosauria, Ornithischia) from the Cenomanian of Texas. Journal of Vertebrate Paleontology 18: 718–738.

Head, J. J. 2001. A reanalysis of the phylogenetic position of *Eolambia caroljonesa* (Dinosauria, Iguanodontia). Journal of Vertebrate Paleontology 21: 392–396.

Heany, L. R. 1978. Island area and body size in insular mammals: evidence from the tri-colored squirrel (*Callosciurus prevosti*) of southeast Asia. Evolution 32: 29–44.

Henderson, D. M. In press. A mathematical and computational model of a quadrupedally walking pterosaur. Zoological Journal of the Linnean Society.

Hennig, W. 1950. *Grundzüge Einer Theorie der Phylogenetischen Systematik*. Deutscher Zentralverlag, Berlin. 370 pp.

Hennig, W. 1965. Phylogenetic systematics. Annual Review of Entomology 10: 97–116.

Hennig, W. 1966. *Phylogenetic Systematics* (translated by Davis, D. D., and Zangerl, R.). University of Illinois Press, Urbana. 263 pp.

Hirayama, R., Brinkman, D. B., and Danilov, I. G. 2000. Distribution and biogeography of non-marine Cretaceous turtles. Russian Journal of Herpetology 7: 181–198.

Hirsch, K. F. 1994. Upper Jurassic eggshells from the western interior of North America. In: Carpenter, K., Hirsch, K. F., and Horner, J. R. (eds.). *Dinosaur Eggs and Babies*. Cambridge University Press, New York. Pp. 137–150.

Hirsch, K. F., and Packard, M. J. 1987. Review of fossil eggs and their shell structure. Scanning Electron Microscopy 1: 383–400.

Hirsch, K. F., and Quinn, B. 1990. Eggs and eggshell fragments from the Upper Cretaceous Two Medicine Formation of Montana. Journal of Vertebrate Paleontology 10: 491–511.

Hoffman, A., and Reif, W.-E. 1988. The methodology of the biological sciences: from an evolutionary biological perspective. Neues Jahrbuch für Geologie und Paläontologie, Abhandlungen 177: 185–211.

Hoffman, A., and Reif, W.-E. 1990. On the study of evolution in species-level lineages in the fossil record: controlled methodological sloppiness. Paläontologische Zeitschrift 64: 673–686.

Holtz, T. R., Jr. 2004. Tyrannosauroidea. In: Weishampel, D. B., Dodson, P., and Osmólska, H. (eds.). *The Dinosauria*, 2nd edition. University of California Press, Berkeley. Pp. 111–136.

Holtz, T. R., Jr., Molnar, R. E., and Currie, P. J. 2004. Basal Tetanurae. In: Weishampel, D. B., Dodson, P., and Osmólska, H. (eds.). *The Dinosauria*, 2nd edition. University of California Press, Berkeley. Pp. 71–110.

Holtz, T. R., Jr., and Osmólska, H. 2004. Saurischia. In: Weishampel, D. B., Dodson, P., and Osmólska, H. (eds.). *The Dinosauria*, 2nd edition. University of California Press, Berkeley. Pp. 21–23.

Hooijer, D. A. 1957. Three new giant rats from Flores, Lesser Sunda Islands. Zoologische Mededeelingen 35: 300–313.

Hooijer, D. A. 1975. Quaternary mammals west and east of Wallace's Line. Netherlands Journal of Zoology 25: 46–56.

Hooker, J. J., Milner, A. C., and Sequeira, S. E. K. 1991 An ornithopod dinosaur from the Late Cretaceous of West Antarctica. Antarctic Science 3: 331–332.

Hooley, R. W. 1914. On the ornithosaurian genus *Ornithocheirus* with a review of the specimens from the Cambridge Greensand in the Sedgwick Museum, Cambridge. Annals and Magazine of Natural History, Series 8, 13: 529–557.

Hope, S. 2002. The Mesozoic radiation of Neornithes. In: Chiappe, L. M., and Witmer, L. M. (eds.). *Mesozoic Birds: Above the Heads of Dinosaurs*. University of California Press, Berkeley. Pp. 337–385.

Horner, J. R. 1984. The nesting behavior of dinosaurs. Scientific American 250 (4): 130–137.

Horner, J. R. 1999. Egg clutches and embryos of two hadrosaurian dinosaurs. Journal of Vertebrate Paleontology 19: 607–611.

Horner, J. R. 2000. Dinosaur reproduction and parenting. Annual Review of Earth and Planetary Sciences 28: 19–45.

Horner, J. R., and Currie, P. J. 1994. Embryonic and neonatal morphology and ontogeny of a new species of *Hypacrosaurus* (Ornithischia, Lambeosauridae) from Montana and Alberta. In: Carpenter, K., Hirsch, K. F., and Horner, J. R. (eds.). *Dinosaur Eggs and Babies*. Cambridge University Press, New York. Pp. 312–336.

Horner, J. R., and Dobb, E. 1997. *Dinosaur Lives*. HarperCollins, New York. 244 pp.

Horner, J. R., and Gorman, J. 1988. *Digging Dinosaurs*. Workman, New York. 210 pp.

Horner, J. R., and Makela, R. 1979. Nest of juveniles provides evidence of family structure among dinosaurs. Nature 282: 296–298.

Horner, J. R., Padian, K., and Ricqlès, A. de. 2001. Comparative osteohistology of some embryonic and perinatal archosaurs: developmental and behavioral implications for dinosaurs. Paleobiology 27: 30–58.

Horner, J. R., Padian, K., and Ricqlès, A. de. 2005. How dinosaurs grew so large—and so small. Scientific American 293: 56–63.

Horner, J. R., Ricqlès, A. de, and Padian, K. 1999. Variation in dinosaur skeletochronology indicators: implications for age assessment and physiology. Paleobiology 25: 295–304.

Horner, J. R., Ricqlès, A. de, and Padian, K. 2000. Long bone histology of the hadrosaurid dinosaur *Maiasaura peeblesorum*: growth dynamics and physiology based on an ontogenetic series of skeletal elements. Journal of Vertebrate Paleontology 20: 115–129.

Horner, J. R., and Weishampel, D. B. 1988. A comparative embryological study of two ornithischian dinosaurs. Nature 332: 256–257.

Horner, J. R., Weishampel, D. B., and Forster, C. A. 2004. Hadrosauridae. In: Weishampel, D. B., Dodson, P., and Osmólska, H. (eds.). *The Dinosauria*, 2nd edition. University of California Press, Berkeley. Pp. 438–463.

Horváth, F. 1974. Application of plate tectonics to the Carpatho-Pannon region; a review. Acta Geologica Academiae Scientiarum Hungaricae 18: 243–255.

Howard, R. S., and Lively, C. M. 2002. The ratchet and the Red Queen: the maintenance of sex in parasites. Journal of Evolutionary Biology 15: 648–656.

Howse, S. C. B., and Milner, A. R. 1993. *Ornithodemus*—a maniraptoran theropod dinosaur from the Lower Cretaceous of the Isle of Wight, England. Palaeontology 36: 425–437.

Huelsenbeck, J. P. 1994. Comparing the stratigraphic record to estimates of phylogeny. Paleobiology 20: 470–483.

Huene, F. von. 1932. *Die Fossile Reptil-Ordnung Saurischia, Ihre Entwicklung und Geschichte*. Monographien für Geologie und Paläontologie, Serie 1, Heft 4, t. 1–2. Gebrüder Borntraeger, Leipzig, Germany. 361 pp.

Huismans, R. S., Bertotti, G., Ciulavu, D., Sanders, C. A. E., Cloetingh, S., and Dinu, C. 1997. Structural evolution of the Transylvanian Basin (Romania): a sedimentary basin in the bend zone of the Carpathians. Tectonophysics 272: 249–268.

Hutchinson, G. E. 1959. Homage to Santa Rosalia, or why are there so many kinds of animals? American Naturalist 93: 117–125.

Huxley, T. H. 1867. On *Acantholpholis horridus*, a new reptile from the Chalk-Marl. Geological Magazine 4: 65–67.

International Committee of Zoological Nomenclature. 1988. Opinion 1483: *Rhabdodon* Matheron, 1869 (Reptilia, Ornithischia) conserved. Bulletin of Zoological Nomenclature 45: 85–86.

Ion, J., Antonescu, E., Melinte, M. C., and Szasz, L. 1988. Biostratigrafia integrată a Santonian-Maastrichtianului din România. Anuarul Institutului Geologic al României 70: 101–107.

Janis, C. 1976. The evolutionary strategy of the Equidae and the origins of rumen and cecal digestion. Evolution 30: 757–774.

Jarman, P. J. 1974. The social organization of antelope in relation to their ecology. Behaviour 48: 215–267.

Ji Q., Currie, P. J., Norell, M., and Ji S.-A. 1998. Two feathered theropods from the Upper Jurassic / Lower Cretaceous strata of northeastern China. Nature 393: 753–761.

Jianu, C. M. 1992. Un fragment de dentar de hadrosaurian (Reptilia, Ornithischia) în Cretacicul superior din Bazinul Hațeg (Sânpetru, Județul Hunedoara). Sargetia 15: 385–396.

Jianu, C. M. 1994. A right dentary of *Rhabdodon priscus* Matheron 1869 (Reptilia: Ornithischia) from the Maastrichtian of Hațeg Basin (Romania). Sargetia 16: 29–35.

Jianu, C.-M., and Boekschoten, G. J. 1999a. The Hațeg area: island or outpost? In: Reumer, J. W. F., and Vos, J. de (eds.). *Elephants Have a Snorkell*. Deinsea Special Vol. 7. Natural History Museum, Rotterdam. Pp. 195–199.

Jianu, C.-M., and Boekschoten, G. J. 1999b. A new look at the Hațeg Island. In: Canudo, J. I., and Cuenca-Bescós, G. (eds.). *4th European Workshop on Vertebrate Palaeontology, Albarracin, Spain: Program and Abstracts*. P. 64.

Jianu, C.-M., and Weishampel, D. B. 1998. Familia Nopcsa și importanța sa istorică în Țara Hațegului. In: *750 de Ani de la Prima Atestare Documentară a Țarii Hațegului 1247-1997, Hațeg, România.* Pp. 202-228.

Jianu, C.-M., and Weishampel, D. B. 1999. The smallest of the largest: a new look at possible dwarfing in sauropod dinosaurs. Geologie en Mijnbouw 78: 335-343.

Jianu, C.-M., Weishampel, D. B., and Știucă, E. 1997. Old and new pterosaur material from the Hațeg Basin (Late Cretaceous) of western Romania, and comments about pterosaur diversity in the Late Cretaceous of Europe. *2nd European Workshop of Vertebrate Paleontology, Quillan, France: Abstracts.* [Unpaginated.]

Jokela, J., Schmidt-Hempel, P., and Rigby, M. C. 2000. Dr. Pangloss restrained by the Red Queen—steps toward a unified defense theory. Oikos 89: 267-274.

Jones, D. S., and Gould, S. J. 1999. Direct measurement of age in fossil *Gryphaea*: the solution to a classic problem in heterochrony. Paleobiology 25: 158-188.

Joyce, W. G. 2007. Phylogenetic relationships of Mesozoic turtles. Bulletin of the Peabody Museum of Natural History 48: 3-102.

Jurcsák, T. 1982. Occurrences nouvelles des sauriens mésozoïques de Roumanie. Vertebrata Hungarica 21: 175-184.

Jurcsák, T., and Kessler, E. 1986. Evoluția avifaunaei pe teritorul României, I. Nymphaea 16: 577-615.

Jurcsák, T., and Kessler, E. 1987. Evoluția avifaunaei pe teritorul României, II: morfologia speciilor fosile. Nymphaea 17: 583-609.

Jurcsák, T., and Kessler, E. 1991. The Lower Cretaceous paleofauna from Cornet, Bihor County, Romania. Nymphaea 21: 5-32.

Jurcsák, T., and Popa, E. 1978. Resturi de dinozaurieni în bauxitele de la Cornet (Bihor): notă preliminară. Nymphaea 6: 61-64.

Jurcsák, T., and Popa, E. 1979. Dinozaurieni ornitopozi din bauxitele de la Cornet (Munții Pădurea Craiului). Nymphaea 7: 37-75.

Jurcsák, T., and Popa, E. 1984. Pterosaurians from the Cretaceous of Cornet, Roumania. In: Reif, W.-E., and Westphal, F. (eds.). *Third Symposium on Mesozoic Terrestrial Ecosystems: Short Papers.* Attempto Verlag, Tübingen, Germany. Pp. 117-118.

Kaneshiro, K. Y. 1989. The dynamics of sexual selection and founder effects in species formation. In: Giddings, L. V., Kaneshiro, K. Y., and Anderson, W. W. (eds.). *Genetics, Speciation, and the Founder Principle.* Oxford University Press, Oxford. Pp. 279-296.

Kawecki, T. J. 1998. Red Queen meets Santa Rosalia: arms races and the evolution of host specialization in organisms with parasitic lifestyles. American Naturalist 152: 635-651.

Kellner, A. W. A. 2003. Pterosaur phylogeny and comments on the evolutionary history of the group. In: Buffetaut, E., and Mazin, J.-M. (eds.). *Evolution and Palaeobiology of Pterosaurs*. Geological Society of London Special Publication 217. Geological Society, London. Pp. 105–137.

Kessler, E. 1984. Lower Cretaceous birds from Cornet (Roumania). In: Reif, W.-E., and Westphal, F. (eds.). *Third Symposium on Mesozoic Terrestrial Ecosystems: Short Papers*. Attempto Verlag, Tübingen, Germany. Pp. 119–121.

Kielan-Jaworowska, Z., Cifelli, R. L., and Luo Z. X. 2004. *Mammals from the Age of Dinosaurs: Origins, Evolution, and Structure*. Columbia University Press, New York. 630 pp.

Kielan-Jaworowska, Z., and Gambaryan, P. P. 1994. Postcranial anatomy and habits of Asian multituberculate mammals. Fossils and Strata 36: 1–92.

Kielan-Jaworowska, Z., and Hurum, J. H. 2001. Phylogeny and systematics of multituberculate mammals. Paleontology 44: 389–429.

Kirkland, J. I. 1998. A new hadrosaurid from the upper Cedar Mountain Formation (Albian–Cenomanian: Cretaceous) of eastern Utah—the oldest known hadrosaurid (lambeosaurine?). In: Lucas, S. G., Kirkland, J. E., and Estep, J. W. (eds.). *Lower and Middle Cretaceous Terrestrial Ecosystems*. New Mexico Museum of Natural History and Science Bulletin 14. New Mexico Museum of Natural History and Science, Albuquerque. Pp. 283–295.

Kleiber, M. 1961. *The Fire of Life*. John Wiley & Sons, New York. 453 pp.

Klein, N., and Sander, M. 2008. Ontogenetic stages in the long bone histology of sauropod dinosaur. *Paleobiology* 34: 247–263.

Klingenberg, C. P. 1998. Heterochrony and allometry: an analysis of evolutionary change in ontogeny. Biological Review 73: 79–123.

Krause, D. W. 1982. Jaw movement, dental function, and diet in the Paleocene multituberculate *Ptilodus*. Paleobiology 8: 265–281.

Krause, D. W., and Jenkins, F. A. 1982. The postcranial skeleton of North American multituberculates. Bulletin of the Museum of Comparative Zoology 150: 199–246.

Lamar, W. W. 1997. *The World's Most Spectacular Reptiles and Amphibians*. World Publications, Tampa, FL. 208 pp.

Lambrecht, K. 1929. Mesozoische und tertiäre Vögelreste aus Siebenbürgen. *Xe Congrès International de Zoologie, Budapest*, part 2. Pp. 1262–1275.

Lapparent, A. F. de. 1947. *Les Dinosauriens du Crétacé Supérieur du Midi de la France*. Mémoires de la Société Géologique de France, Nouvelle Série 56. Société Géologique de France, Paris. 54 pp.

Lapparent de Broin, F. de, and Murelaga, X. 1996. Une nouvelle faune de chéloniens dans le Crétacé supérieur européen. Comptes Rendus de l'Académie des Sciences 323: 729–735.

Lapparent de Broin, F. de, and Murelaga, X. 1999. Turtles from the Upper Cretaceous of Laño (Iberian Peninsula). Estudios del Museo de Ciencias Naturales de Alva 14: 135–211.

Lapparent de Broin, F., Murelga-Bereikua, X., and Codrea, V. 2004. Presence of Dortokidae (Chelonii, Pleurodira) in the earliest Tertiary of the Jibou Formation: paleobiogeographical implications. Acta Palaeontologica Romaniae 4: 203–215.

Laurent, Y., Le Loeuff, J., and Buffetaut, E. 1997. Les Hadrosauridae (Dinosauria, Ornithopoda) du Maastrichtien supérieur des Corbières orientales (Aude, France). Revue de Paléobiologie 16: 411–423.

Le Grand, H. E. 1988. *Drifting Continents and Shifting Theories*. Cambridge University Press, Cambridge. 313 pp.

Le Loeuff, J. 1991. The Campano-Maastrichtian vertebrate faunas of southern Europe and their relationship with other faunas in the world: palaeobiogeographical implications. Cretaceous Research 12: 93–114.

Le Loeuff, J. 1993. European titanosaurids. Revue de Paléobiologie, Vol. Spécial 7: 105–117.

Le Loeuff, J. 1995. *Ampelosaurus atacis* (nov. gen., nov. sp.), et nouveau Titanosauridae (Dinosauria, Sauropoda) du Crétacé supérieur de la Haute Vallée de l'Aude (France). Comptes Rendus de l'Académie des Sciences de Paris, Série 2, 321: 693–699.

Le Loeuff, J. 1997. Musée des Dinosaures, Espéraza, Aude, France. In: Currie, P. J., and Padian, K. (eds.). *Encyclopedia of Dinosaurs*. Academic Press, New York. Pp. 454–455.

Le Loeuff, J. 2002. Romanian Late Cretaceous dinosaurs: big dwarfs or small giants? In: *7th European Workshop of Vertebrate Palaeontology, Sibiu, Romania: Abstracts Volume and Excursions Field Guide*. P. 31.

Le Loeuff, J. 2005. Osteology of *Ampelosaurus atacis* (Titanosauria) from southern France: In: Tidwell, V., and Carpenter, K (eds.). *Thunder-Lizards: The Sauropodomorph Dinosaurs*. Indiana University Press, Bloomington. Pp. 115–137.

Le Loeuff, J., and Buffetaut, E. 1990. *Tarascosaurus salluvicus* nov. gen., nov. sp., dinosaure Théropode du Crétacé supérieur du sud de la France. Geobios 25: 585–594.

Le Loeuff, J., and Buffetaut, E. 1995. The evolution of Late Cretaceous non-marine vertebrate fauna in Europe. In: Sun A.-L., and Wang Y.-Q. (eds.). *Sixth Symposium on Mesozoic Terrestrial Ecosystems and Biotas: Short Papers*. China Ocean Press, Beijing. Pp. 181–184.

Le Loeuff, J., and Buffetaut, E. 1998. A new dromaeosaurid theropod from the Upper Cretaceous of southern France. Oryctos 1: 105–112.

Le Loeuff, J., Buffetaut, E., Cavin, L., Martin, M., Martin, V., and Tong H. 1994. An armoured titanosaurid sauropod from the Late Cretaceous of southern France and the occurrence of osteoderms in the Titanosauridae. Gaia 10: 155–159.

Le Loeuff, J., and Moreno, C. 1992. *Les Dinosaures de l'Aude*. Euro Impression, Couiza, France. 16 pp.

Le Loeuff, J., and Souillat, C. 1997. Les dinosaures des Pyrénées. Pyrénées Magazine 54: 44–55.

Lillegraven, J. A., Kielan-Jaworowska, Z., and Clemens, W. A. 1979. *Mesozoic Mammals*. University of California Press, Berkeley. 311 pp.

Lindfors, S. M., Csiki, Z., Grigorescu, D., and Friis, E. M. 2010. Preliminary account of plant mesofossils from the Maastrichtian Budurone microvertebrate site of the Haţeg Basin, Romania. Palaeogeography, Palaeoclimatology, Palaeoeology 293: 353–359.

Lindgren, J., Currie, P. J., Siverson, M., Rees, J., Cederstrom, P., and Lindgren, F. 2008. The first neoceratopsian dinosaur remains in Europe. Palaeontology 50: 929–937.

Lindstedt, S. L., and Boyce, M. S. 1985. Seasonality, fasting endurance, and body size in mammals. American Naturalist 125: 873–878.

Lindstedt, S. L., and Calder, W. A. 1981. Body size, physiological time, and longevity of homeothermic animals. Quarterly Review of Biology 56: 1–16.

Linzer, H.-G., Frisch, W., Zweigel, P., Girbacea, R., Hann, H.-P., and Moser, F. 1998. Kinematic evolution of the Romanian Carpathians. Tectonophysics 297: 133–156.

Lively, C. M. 1996. Host-parasite coevolution and sex. Bioscience 46: 107–114.

Lloyd, C. R. 1982. The mid-Cretaceous Earth: paleogeography, ocean circulation and temperature, atmospheric circulation. Journal of Geology 90: 393–413.

Lomolino, M. V. 1984. Immigrant selection, predation, and the distributions of *Microtus pennsylvanicus* and *Blarina brevicauda* on islands. American Naturalist 123: 468–483.

Lomolino, M. V. 1985. Body size of mammals on islands: the Island Rule re-examined. American Naturalist 125: 310–316.

Lomolino, M. V. 1986. Mammalian community structure on islands: the importance of immigration, extinction, and interactive effects. Biological Journal of the Linnean Society 28: 1–21.

Long, J. A., and McNamara, K. J. 1995. Heterochrony in dinosaur evolution. In: McNamara, K. J. (ed.). *Evolutionary Change and Heterochrony*. John Wiley & Sons, New York. Pp. 151–168.

López-Martínez, N., Canudo, J. I., Ardèvol, L., Pereda-Suberbiola, X., Orue-Etxebarria, X., Cuenca-Bescós, G., Ruiz-Omeñaca, J. I., Murelaga, X., and

Feist, M. 2001. New dinosaur sites correlated with upper Maastrichtian pelagic deposits in the Spanish Pyrenees: implications for the dinosaur extinction pattern in Europe. Cretaceous Research 22: 41–61.

Lovaszi, P., and Moskat, C. 2004. Break-down of arms race between the red-backed shrike (*Lanius collurio*) and common cuckoo (*Cuculus canorus*). Behaviour 141: 245–262.

Lucas, S. G., and Sullivan, R. M. 2000. The sauropod dinosaur *Alamosaurus* from the Upper Cretaceous of the San Juan Basin, New Mexico. New Mexico Museum of Natural History Bulletin 17: 147–156.

Lull, R. S., and Wright, N. E. 1942. Hadrosaurian dinosaurs of North America. Geological Society of America Special Papers 40. Geological Society of America, New York. 242 pp.

Lydekker, R. 1877. Notices of new and other Vertebrata from Indian Tertiary and Secondary rocks. Records of the Geological Survey of India 10: 30–42.

Lydekker, R. 1888. *The Orders Ornithosauria, Crocodilia, Dinosauria, Squamata, Rhynchocephalia, and Proterosauria*, part 1 of *Catalogue of the Fossil Reptilia and Amphibia in the British Museum (Natural History)*. British Museum (Natural History), London. 309 pp.

MacArthur, R. W., and Wilson, E. O. 1967. *The Theory of Island Biogeography*. Princeton University Press, Princeton, NJ. 203 pp.

Machalski, M., Jagt, J. W. M., Dorstangs, R. W., Mulder, E. W. A., and Radwañski, A. 2003. Campanian and Maastrichtian mosasaurid reptiles from central Poland. Acta Palaeontologica Polonica 48: 397–408.

Maio D. 1988. *Skull Morphology of* Lambdopsalis bulla *(Mammalia, Multituberculata) and Its Implications to Mammalian Evolution*. Contributions to Geology, University of Wyoming, Special Paper 4. 104 pp.

Maiorana, V. J. 1973. Evolutionary strategies and body size in a guild of mammals. In: Damuth, J., and MacFadden, B. J. (eds.). *Body Size in Mammalian Paleobiology*. Cambridge University Press, New York. Pp. 69–102.

Makovicky, P. J., Kobayashi, Y., and Currie, P. J. 2004. Ornithomimosauria. In: Weishampel, D. B., Dodson, P., and Osmólska, H. (eds.). *The Dinosauria*, 2nd edition. University of California Press, Berkeley. Pp. 137–150.

Makovicky, P. J., and Norell, M. A. 2004. Troodontidae. In: Weishampel, D. B., Dodson, P., and Osmólska, H. (eds.). *The Dinosauria*, 2nd edition. University of California Press, Berkeley. Pp. 184–195.

Mamulea, M. A. 1953. Studii geologice în regiunea Sânpetru-Pui (Bazinul Haţegului). Anuarul Institutului Geologic al României 10: 301–333.

Mantell, G. A. 1825. Notice on the *Iguanodon*, a newly discovered fossil reptile, from the Sandstone of Tilgate Forest, in Sussex. Philosophical Transactions of the Royal Society of London 115: 179–186.

Mantell, G. A. 1832. *Geology of the South East of England*. London. 415 pp.

Mărgărit, C., and Mărgărit, M. 1962. Asupra prezenței unor resturi de plante fosile în împrejurimile localității Densuș (Bazinul Hațeg). Studii și Cercetări de Geologie, Geofizică, Geografie, Serie Geologie 12: 471–476.

Marinescu, F. 1989. Lentila de bauxită 204 de la Brusturi-Cornet (Jud. Bihor), zăcămînt fosilifer cu dinozauri. Ocrotirea Naturii și a Mediului Înconjurător 33: 125–132.

Marshall, L. G., and Corruccini, R. S. 1978. Variability, evolutionary rates, and allometry in dwarfing lineages. Paleobiology 4: 101–119.

Martens, K., and Schön, I. 2000. Parasites, predators, and the Red Queen. Trends in Ecology and Evolution. 15: 392–393.

Martin, J. E., and Buffetaut, E. 2008. *Crocodylus affuvelensis* Matheron, 1869 from the Late Cretaceous of southern France: a reassessment. Zoological Journal of the Linnean Society 152: 567–580.

Martin, J. E., Csiki, Z., Grigorescu, D., and Buffetut, E. 2006. Late Cretaceous crocodilian diversity in Hațeg Basin, Romania. Hantkeniana 5: 31–37.

Martin, J. E., and Delfino, M. 2010. Recent advances in the comprehension of the biogeography of Cretaceous European eusuchians. Palaeogeography, Palaeoclimatology, Palaeoecology 293: 406–418.

Martin, J. E., Rabi, M., and Csiki, Z. 2010. Survival of *Theriosuchus* (Mesoeucrocodylia: Atoposauridae) in a Late Cretaceous archipelago: a new species from the Maastrichtian of Romania. Naturwissenschaften 97: 845–854.

Martin, L. D. 1983. The origin and early radiation of birds. In: Brush, A. H., and Clark, G. A. (eds.). *Perspectives in Ornithology*. Cambridge University Press, Cambridge. Pp. 291–338.

Matheron, P. 1869a. Notice sur les reptiles fossiles des dépôts fluvio-lacustres crétacés du bassin à lignite de Fuveau. Mémoires de l'Académie Impériale des Sciences, Belles-Lettres, et [Beaux-]Arts de Marseille 1868–1869: 345–379.

Matheron, P. 1869b. Note sur les reptiles fossiles des dépôts fluvio-lacustres crétacés du bassin à lignite de Fuveau. Bulletin de la Société Géologique de France 26: 781–795.

Matthews, J. 1998. *The Battle of Brazil*. Applause Publications, New York. 272 pp.

Mayor, A. 2000. *The First Fossil Hunters: Paleontology in Greek and Roman Times*. Princeton University Press, Princeton, NJ. 361 pp.

Mayr, E. 1942. *Systematics and the Origin of Species*. Columbia University Press, New York. 334 pp.

Mayr, E. 1963. *Animal Species and Evolution*. Columbia University Press, New York. 797 pp.

Mayr, E. 1965. The nature of colonization in birds. In: Baker, H. G., and Stebbins,

G. L. (eds.). *The Genetics of Colonizing Species*. Academic Press, New York. Pp. 29–47.

McGowan, G. J. 2002. Albanerpetontid amphibians from the Lower Cretaceous of Spain and Italy: a description and reconsideration of their systematics. Zoological Journal of the Linnean Society 135: 1–32.

McGowan, G. J., and Evans, S. E. 1995. Albanerpetontid amphibians from the Cretaceous of Spain. Nature 373: 143–145.

McKenna, M. C., and Bell, S. K. 2000. *Classification of Mammals*. Columbia University Press, New York. 640 pp.

McKinney, M. L. 1984. Allometry and heterochrony in an Eocene echinoid lineage: morphological change as a byproduct of size selection. Paleobiology 10: 407–419.

McKinney, M. L. 1986. Ecological causation of heterochrony: a test and implications for evolutionary theory. Paleobiology 12: 282–289.

McKinney, M. L. 1988. Classifying heterochrony: allometry, size, and time. In: McKinney, M. L. (ed.). *Heterochrony in Evolution*. Plenum, New York. Pp. 17–34.

McKinney, M. L. 1999. Heterochrony: beyond words. Paleobiology 25: 149–153.

McKinney, M. L., and McNamara, K. J. 1991. *Heterochrony: The Evolution of Ontogeny*. Plenum, New York. 437 pp.

McKinney, M. L., McNamara, K. J., and Zachos, L. G. 1990. Heterochronic hierarchies: application and theory in evolution. Historical Biology 3: 269–287.

McKinney, M. L., and Schoch, R. M. 1985. Titanothere allometry, heterochrony, and biomechanics: revisiting an evolutionary classic. Evolution 39: 1352–1363.

McNamara, K. J. 1986. A guide to the nomenclature of heterochrony. Journal of Paleontology 60: 4–13.

McNamara, K. J. 1988. Patterns of heterochrony in the fossil record. Trends in Ecology and Evolution 3: 176–180.

McNamara, K. J. 1997. *Shapes of Time*. Johns Hopkins University Press, Baltimore. 342 pp.

McNamara, K. J., and Trewin, N. H. 1993. A euthycarcinoid arthropod from the Silurian of Western Africa. Palaeontology 36: 319–335.

McNaughton, S. J. 1975. r- and K-selection in *Typha*. American Naturalist 109: 251–263.

Meers, M. B. 1999. Evolution of the crocodylian forelimb: anatomy, biomechanics, and functional morphology. Ph.D. diss., Johns Hopkins University School of Medicine, Baltimore, Maryland.

Mikhailov, K. E. 1991. Classification of fossil eggshells of amniote vertebrates. Acta Palaeontologica Polonica 36: 193–238.

Mikhailov, K. E., Bray, E. S., and Hirsch, K. R. 1996. Parataxonomy of fossil egg remains (Vertebrata): principles and applications. Journal of Vertebrate Paleontology 16: 763–769.

Miller, H. V. 1934. *Tropic of Cancer*. Grove Press, New York. 318 pp.

Milner, A. C., Hooker, J. J., and Sequeira, S. E. K. 1992. An ornithopod dinosaur from the Upper Cretaceous of the Antarctic Peninsula. Journal of Vertebrate Paleontology 12 (Suppl.): 44A.

Mindszenty, A., D'Argenio, B., and Aiello, G. 1995. Lithospheric bulges recorded by regional unconformities: the case of Mesozoic–Tertiary Apulia. Tectonophysics 252: 137–161.

Minkovska, V., Peybernès, B., and Nikolov, T. 2002. Palaeogeography and geodynamic evolution of the Balkanides and Moesian "microplate" (Bulgaria) during the earliest Cretaceous. Cretaceous Research 23: 37–48.

Mohr, C. O. 1943. Cattle droppings as ecological units. Ecological Monographs 13: 275–298.

Morey, D. F. 1994. The early evolution of the domestic dog. American Scientist 82: 217–247.

Mulder, E. W. A. 1984. Resten van *Telmatosaurus* (Ornithischia, Hadrosauridae) uit het Boven-Krijt van Zuid-Limburg. Grondboor en Hamer 1984: 108–115.

Mulder, E. W. A., Kuypers, M. M. M., Jagt, J. W. M., and Peeters, H. H. G. 1997. A new late Maastrichtian hadrosaurid dinosaur record from northeast Belgium. Neues Jahrbuch für Geologie und Paläontologie Monatshefte 1997: 339–347.

Naish, D., and Dyke, G. J. 2004. *Heptasteornis* was no ornithomimid, troodontid, dromaeosaurid, or owl: the first alvarezsaurid (Dinosauria: Theropoda) from Europe. Neues Jahrbuch für Geologie und Paläontologie Monatshefte 2004: 385–401.

Nelson, G., and Platnick, N. 1981. *Systematics and Biogeography: Cladistics and Vicariance*. Columbia University Press, New York. 567 pp.

Nelson, G., and Rosen, D. E. (eds.). 1981. *Vicariance Biogeography*. Columbia University Press, New York. 593 pp.

Nessov, L. A. 1993. First central Asian stegosaur. Dinosaur Report, Winter 1993: 3.

Nessov, L. A. 1995. *Dinozavri Severnoid Evrazii: Novye Dannye o Sostave Kompleksov, Ekologii, i Paleobiogeografii* [*Dinosaurs of Northern Eurasia: New Data on the Composition of Assemblages, Ecology, and Paleobiogeography*]. Sankt-Petersburg University, St. Petersburg. 156 pp. [In Russian.]

Nessov, L. A., and Yarkov, A. A. 1993. [Hesperornithes in Russia.] Russkii Ornithologicheskii Zhurnal 2: 37–54. [In Russian.]

Newton, E. T. 1892. Note on an iguanodont tooth from the Lower Chalk ("Totternhoe stone"), near Hitchin. Geological Magazine, Series 3, 9: 49–50.

Nicolescu, S. 1998. Glossary of geological localities in the former Austro-Hungarian Empire, now in Romania. Canadian Mineralogist 36: 1373–1381.

Nikolov, I., and Westphal, F. 1976. Mosasaurier-Funde aus der Oberkreide von Nordwest-Bulgarien. Neues Jahrbuch für Geologie und Paläontologie Monatshefte 1976: 608–613.

Nopcsa, F. 1897. Vorläufiger Bericht über das Auftreten von oberer Kreide im Hátszeger Tale in Siebenbürgen. Verhandlungen der Königlich-Kaiserlichen Geologischen Reichsanstalt 1897: 374–374.

Nopcsa, F. 1900. Dinosaurierreste aus Siebenbürgen: Schädel von *Limnosaurus transsylvanicus* nov. gen. et spec. Denkschriften der Kaiserlichen Akademie der Wissenschaften Wien, Mathematisch-Naturwissenschaftliche Classe 68: 555–591.

Nopcsa, F. 1902a. Über das Vorkommen von Dinosauriern be Szentpéterfalva. Zeitschrift der Deutschen Geologischen Gesellschaft, Monatsbericht 1902: 34–39.

Nopcsa, F. 1902b. Notizen über cretacische Dinosaurier. Sitzungsberichte der Königlichen Akademie der Wissenschaften Wien 111: 93–114.

Nopcsa, F. 1903. Neues über *Compsognathus*. Neues Jahrbuch für Mineralogie, Geologie, und Palaeontologie, Beilageband 16: 476–494.

Nopcsa, F. 1904. Dinosaurierreste aus Siebenbürgen, III: weitere Schädelreste von *Mochlodon*. Denkschriften der Kaiserlichen Akademie der Wissenschaften Wien, Mathematisch-Naturwissenschaftliche Classe 74: 229–264.

Nopcsa, F. 1905a. Remarks on the supposed clavicle of the sauropodous dinosaur *Diplodocus*. Proceedings of the Zoological Society of London 1905: 241–250.

Nopcsa, F. 1905b. Zur Geologie der Gegend zwischen Gyulafehérvár, Déva, Ruszkabánya, und der rumänischen Landesgrenze. Mitteilungen aus dem Jahrbuch der Königlichen Ungarischen Geologischen Anstalt 14: 93–279.

Nopcsa, F. 1905c. Zur geologie von Nordalbanien. Jahrbuch der Königlich-Kaiserlichen Geologischen Reichsanstalt 55: 85–152.

Nopcsa, F. 1908. Die Mineralquellen Makedoniens. Mitteilungen der Königlich-Kaiserlichen Geographischen Gesellschaft 1908: 242–292.

Nopcsa, F. 1910. *Aus Sala und Klementi*. D. A. Kajon, Sarajevo. 115 pp.

Nopcsa, F. 1911. Zur Stratigraphie und Tektonic des Vilajets Skutari in Nordalbanien. Jahrbuch der Königlich-Kaiserlichen Geologischen Reichsanstalt 61: 229–284.

Nopcsa, F. 1912. *Haus und Hausrat im Katholischen Nordalbanien*. B. H. Instituts für Balkanforschung, Bosnia-Herzegovina Landesdruckerei, Sarajevo. 90 pp.

Nopcsa, F. 1914a. Die Lebensbedingungen der Dinosaurier Siebenbürgens. Centralblatt fur Mineralogie, Geologie, und Palaeontologie 1914: 564–574.

Nopcsa, F. 1914b. Über das Vorkommen der Dinosaurier in Siebenbürgen. Verhandlungen der Königlich-Kaiserlichen Zoologisch-Botanischen Gesellschaft 54: 12–14.

Nopcsa, F. 1915. Die Dinosaurier der siebenbürgischen Landesteile Ungarns. Mitteilungen aus dem Jahrbuch der Königlischen Ungarischen Geologischen Reischsanstalt 23: 1–27.

Nopcsa, F. 1916. Zur Geschichte der Kartographie Nordalbaniens. Mitteilungen der Königlich-Kaiserlichen Geographischen Gesellschaft 1916: 520–585.

Nopcsa, F. 1917a. Über Dinosaurier, II: die Riesenformen unter den Dinosauriern. Centralblatt für Mineralogie, Geologie, und Palaeontologie 1917: 332–348.

Nopcsa, F. 1917b. Über Dinosaurier, III: über die Pubis der Orthopoden. Centralblatt für Mineralogie, Geologie, und Palaeontologie 1917: 348–351.

Nopcsa, F. 1917c. A Dinosaurusok élete és szerepe. Pótfüzetek a Természettudományi Közlönyhöz 127–128: 113–164.

Nopcsa, F. 1918. Neues über Geschlechtsunterschiede bei Orthopoden. Centralblatt für Mineralogie, Geologie, und Palaeontologie 1918: 186–198.

Nopcsa, F. 1921. Geologische Grundzüge der Dinariden. Geologische Rundschau 12: 1–19.

Nopcsa, F. 1922. On the probable habits of the dinosaur *Struthiomimus*. Annals and Magazine of Natural History 9: 152–155.

Nopcsa, F. 1923a. Dinosaurierreste aus Siebenbürgen, IV: die Wirbelsäule von *Rhabdodon* und *Orthomerus*. Palaeontologia Hungarica 1: 273–304.

Nopcsa, F. 1923b. On the geological importance of the primitive reptilian fauna in the Uppermost Cretaceous of Hungary: with a description of a new tortoise (*Kallokibotion*). Quarterly Journal of the Geological Society 29: 100–116.

Nopcsa, F. 1923c. *Kallokibotium* [sic], a primitive amphichelydean tortoise from the Uppermost Cretaceous of Hungary. Palaeontologia Hungarica 1: 1–34.

Nopcsa, F. 1924. The term "arrostic." Science 59: 238–239.

Nopcsa, F. 1925a. Zur Geologie der Kustenkette Nordalbaniens. Mitteilungen aus dem Jahrbuch Königlichen Ungarischen Geologischen Anstalt 24: 133–164.

Nopcsa, F. 1925b. *Albanien: Bauten, Trachten, und Geräte Nordalbaniens*. W. de Gruyter, Berlin. 257 pp.

Nopcsa, F. 1926a. Heredity and evolution. Proceedings of the Zoological Society of London 1926: 633–665.

Nopcsa, F. 1926b. Reversibilität und Dollo'sches Gesetz. Palaeontologische Zeitschrift 7: 184.

Nopcsa, F. 1926c. Die Reptilien der Gosau in neuer Beleuchtung. Centralblatt für Mineralogie, Geologie, und Palaeontologie 1926: 520–523.

Nopcsa, F. 1926d. A 30-million-year-old dinosaur for London: a reconstruction—a 4000-lb. insectivorous dinosaur. Illustrated London News, 11 September 1926.

Nopcsa, F. 1927. Beiträge zur Verteilung der Eruptivgestein. Földtani Közlöny 56: 149–160.

Nopcsa, F. 1928a. Festrede: Gehalten anlasslich des Besuches der Palaeontologischen Gesellschaft in Budapest am 27 September. Geologische Anstalt, Budapest 1928: 1–15.

Nopcsa, F. 1928b. *Palaeontological Notes on Reptiles*. Geologica Hungarica, Series Palaeontologica 1, tome 1, fascicle 1. Edidit Institutum Regni Hungariae Geologicum, Budapest. 84 pp.

Nopcsa, F. 1928c. Paleontological notes on Reptilia, 7: classification of the Crocodilia. In: *Palaeontological Notes on Reptiles*. Geologica Hungarica, Series Palaeontologica 1, tome 1, fascicle 1. Edidit Institutum Regni Hungariae Geologicum, Budapest. Pp. 75–84.

Nopcsa, F. 1929a. *Struthiosaurus transylvanicus*. In: *Dinosaurierreste aus Siebenbürgen, V*. Geologica Hungarica, Series Palaeontologica, fascicle 4. Edidit Institutum Regni Hungariae Geologicum, Budapest. Pp. 247–279.

Nopcsa, F. 1929b. Sexual differences in ornithopodous dinosaurs. Palaeobiologica 2: 187–201.

Nopcsa, F. 1929c. *Geographie und Geologie Nordalbaniens, mit einem Anhang von H. v. Mžik: Beiträge zur Kartographie Albaniens nach Orientalischen Quellen*. Geologica Hungarica, Series Geologica, tome 3. Institutum Regni Hungariae Geologicum, Budapest. 704 pp.

Nopcsa, F. 1930. Zur Systematik und Biologie der Sauropoden. Palaeobiologica 3: 40–52.

Nopcsa, F. 1932. Zur Geschichte der Adria: eine tektonische Studie. Zeitschrift der Deutschen Geologischen Gesellschaft 84: 280–319.

Nopcsa, F. 1934. The influence of geological and climatological factors on the distribution of non-marine fossil reptiles and Stegocephalia. Quarterly Journal of the Geological Society of London 90: 76–140.

Nopcsa, F. 2001. *Reisen in den Balkan* (edited by Elsie, R.). Dukagjini, Peja, Serbia. 527 pp.

Nopcsa, F., and Heidsieck, E. 1933. On the histology of the ribs in immature and half-grown trachodont dinosaurs. Proceedings of the Zoological Society of London 1933: 221–226.

Norell, M. A. 1992. Taxic origin and temporal diversity: the effect of phylogeny. In: Novacek, M. J., and Wheeler, Q. D. (eds.). *Extinction and Phylogeny*. Columbia University Press, New York. Pp. 89–118.

Norell, M. A., Clark, J. M., Chiappe, L. M., and Dashzeveg, D. 1995. A nesting dinosaur. Science 378: 774–776.

Norell, M. A., Clark, J. M., Dashzeveg, D., Barsbold, R., Chiappe, L. M., Davidson, A. R., McKenna, M. C., Perle, A., and Novacek, M. J. 1994. A theropod embryo and the affinities of the Flaming Cliffs dinosaur eggs. Science 266: 779–782.

Norell, M. A., and Makovicky, P. J. 2004. Dromaeosauridae. In: Weishampel, D. B., Dodson, P., and Osmólska, H. (eds.). *The Dinosauria*, 2nd edition. University of California Press, Berkeley. Pp. 196–209.

Norell, M. A., and Novacek, M. J. 1992a. The fossil record and evolution: comparing cladistic and paleontologic evidence for vertebrate history. Science 255: 1690–1693.

Norell, M. A., and Novacek, M. J. 1992b. Congruence between superpositional and phylogenetic patterns: comparing cladistic patterns with fossil records. Cladistics 8: 319–337.

Norman, D. B. 1980. *On the Ornithischian Dinosaur Iguanodon bernissartensis from the Lower Cretaceous of Bernissart (Belgium)*. Mémoires de l'Institut Royal des Sciences Naturelles de Belgique 178. Institut Royal des Sciences Naturelles de Belgique, Brussels. 102 pp.

Norman, D. B. 1984. On the cranial morphology and evolution of ornithopod dinosaurs. In: Ferguson, M. W. J. (ed.). *The Structure, Development, and Evolution of Reptiles*. Symposia of the Zoological Society of London 52. Academic Press, London. Pp. 521–547.

Norman, D. B. 1986. On the anatomy of *Iguanodon atherfieldensis* (Ornithischia: Ornithopoda). Bulletin de l'Institut Royal des Sciences Naturelles de Belgique 56: 281–372.

Norman, D. B. 2004. Basal Iguanodontia. In: Weishampel, D. B., Dodson, P., and Osmólska, H. (eds.). *The Dinosauria*, 2nd edition. University of California Press, Berkeley. Pp. 413–437.

Norman, D. B., Sues, H.-D., Witmer, L. M., and Coria, R. A. 2004. Basal Ornithopoda. In: Weishampel, D. B., Dodson, P., and Osmólska, H. (eds.). *The Dinosauria*, 2nd edition. University of California Press, Berkeley. Pp. 393–412.

Norman, D. B., and Weishampel, D. B. 1985. Ornithopod feeding mechanisms: their bearing on the evolution of herbivory. American Naturalist 126: 151–164.

Norman, D. B., and Weishampel, D. B. 1987. Vegetarian dinosaurs chew it differently. New Scientist 114: 42–45.

Norman, D. B., and Weishampel, D. B. 1991. Feeding mechanisms in some small herbivorous dinosaurs: processes and patterns. In: Rayner, J. M. V., and Wootton, R. J. (eds.). *Biomechanics in Evolution*. Cambridge University Press, Cambridge. Pp. 161–182.

Novacek, M. J., and Norell, M. A. 1982. Fossils, phylogenies, and taxonomic rates of evolution. Systematic Zoology 31: 366–375.

Novacek, M. J., Wyss, A. R., and McKenna, M. C. 1988. The major groups of eutherian mammals. In: Benton, M. J. (ed.). *Mammals*, vol. 2 of *The Phylogeny and Classification of the Tetrapods*. Systematics Association Special Vol. 35B. Clarendon Press, Oxford. Pp. 31–71.

Olson, S. L. 1985. The fossil record of birds. Avian Biology 8: 79–174.

Ősi, A. 2003. A Magyarországi Késö-Kréta dinoszaurusz fauna Vizgálata. Master's thesis, Eötvös Loránd University, Budapest. 66 pp.

Ősi, A. 2004. The first dinosaur remains from the Upper Cretaceous of Hungary (Csehbánya Formation, Bakony Mts). Geobios 37: 749–753.

Ősi, A. 2005. *Hungarosaurus tormai*, a new ankylosaur (Dinosauria) from the Upper Cretaceous of Hungary. Journal of Vertebrate Paleontology 25: 370–383.

Ősi, A., Butler, R. J., and Weishampel, D. B. 2010. A Late Cretaceous ceratopsian dinosaur from Europe with Asian affinities. Nature 465: 466–468.

Ősi, A., Clark, J. M., and Weishampel, D. B. 2007. First report of a new basal eusuchian crocodyliform with multicusped teeth from the Upper Cretaceous (Santonian) of Hungary. Neue Jahrbuch für Geologie und Paläontologie, Abhandlungen 243: 169–177.

Ősi, A., and Főzy, I. 2009. A maniraptoran (Theropoda, Dinosauria) sacrum from the Upper Cretaceous of the Hațeg Basin (Romanian): in search of the lost pterosaurs of Franz Baron Nopcsa. Neues Jahrbuch für Geologie und Paläontologie, Abhandlungen 246: 173–181.

Ősi, A., Jianu, C.-M., and Weishampel, D. B. 2003. Dinosaurs from the Upper Cretaceous of Hungary. In: Petculescu, A., and Știucă, E. (eds.). *Advances in Vertebrate Paleontology: Hen to Panta*. Institute of Speleology "Emil Racoviță," Romanian Academy of Science, Bucharest. Pp. 117–120.

Ősi, A., and Makádi, L. 2009. New remains of Hungarosaurus tormai (Ankylosauria, Dinosauria) from the Upper Cretaceous of Hungary: skeletal reconstruction and body mass estimation. Paläontologische Zeitschrift 83: 227–245. Ősi, A., Weishampel, D. B., and Jianu, C.-M. 2005. First evidence of azhdarchid pterosaurs from the Late Cretaceous of Hungary. Acta Palaeontologica Polonica 50: 777–787.

Osmólska, H., and Barsbold, R. 1990. Troodontidae. In: Weishampel, D. B., Dodson, P., and Osmólska, H. (eds.). *The Dinosauria*. University of California Press, Berkeley. Pp. 259–268.

Osmólska, H., Currie, P. J., and Barsbold, R. 2004. Oviraptorosauria. In: Weishampel, D. B., Dodson, P., and Osmólska, H. (eds.). *The Dinosauria*, 2nd edition. University of California Press, Berkeley. Pp. 165–193.

Ostrom, J. H. 1961. Cranial morphology of the hadrosaurian dinosaurs of North America. Bulletin of the American Museum of Natural History 122: 33–186.

Ostrom, J. H. 1964. A reconsideration of the paleoecology of hadrosaurian dinosaurs. American Journal of Science 262: 975–997.

Ostrom, J. H. 1969. Osteology of *Deinonychus antirrhopus*, an Unusual Theropod from the Lower Cretaceous of Montana. Peabody Museum of Natural History Bulletin 30. Peabody Museum of Natural History, Yale University, New Haven, CT. 165 pp.

Ostrom, J. H. 1980. The evidence for endothermy in dinosaurs. In: Thomas, R. D. K., and Olson, E. C. (eds.). *A Cold Look at the Warm-Blooded Dinosaurs*. Westview Press, Boulder, CO. Pp. 15–54.

Ostrom, J. H. 1990. Dromaeosauridae. In: Weishampel, D. B., Dodson, P., and Osmólska, H. (eds.). *The Dinosauria*. University of California Press, Berkeley. Pp. 268–279.

Otte, D., and Endler, J. A. (eds.). 1989. *Speciation and Its Consequences*. Sinauer Associates, Sunderland, MA. 679 pp.

Owen, R. S. 1842. Report on British fossil reptiles, part II. Report of the Meeting of the British Association for the Advancement of Science 1841: 60–204.

Owen, R. S. 1894. *The Life of Richard Owen*. 2 vols. J. Murray, London. 409 pp., 393 pp.

Padian, K., and Chiappe, L. 1998. The origin of birds and their flight. Scientific American 278: 38–47.

Padian, K., and Horner, J. R. 2004. Dinosaur physiology. In: Weishampel, D. B., Dodson, P., and Osmólska, H. (eds.). *The Dinosauria*, 2nd edition. University of California Press, Berkeley. Pp. 660–671.

Padian, K., Horner, J. R., and Ricqlès, A. de. 2004. Growth in small dinosaurs and pterosaurs: the evolution of archosaurian growth strategies. Journal of Vertebrate Paleontology 24: 555–571.

Pană, I., Grigorescu, D., Csiki, Z., and Costea, C. 2001. Paleo-ecological significance of the continental gastropod assemblages from the Maastrichtian dinosaur beds of the Hațeg Basin. Acta Palaeontologica Romaniae 3: 337–343.

Panaiotu, C. G., and Panaiotu, C. E. 2002. Palaeomagnetic studies. In: *7th European Workshop of Vertebrate Palaeontology, Sibiu, Romania: Abstracts Volume and Excursions Field Guide*. P. 61.

Panaiotu, C. G., and Panaiotu, C. E. 2010. Palaeomagnetism of the Upper Cretaceous Sânpetru Formation (Hațeg Basin, South Carpathians). Palaeogeography, Palaeoclimatology, Palaeoecology 293: 343–352.

Pătrașcu, Ș., Bleahu, M., Panaiotu, C., and Panaiotu, C. E. 1992. The paleomagnetism of the Upper Cretaceous magmatic rocks in the Banat area of South Carpathians: tectonic implications. Tectonophysics 213: 341–352.

Pătrașcu, Ș., and Panaiotu, C. 1990. Paleomagnetism of some Upper Cretaceous deposits in the South Carpathians. Revue Roumaine de Géophysique 34: 67–77.

Pătrașcu, Ș., Panaiotu, C., Șeclăman, M., and Panaiotu, C. E. 1994. Timing of rotational motion of Apuseni Mountains (Romania): paleomagnetic data from Tertiary magmatic rocks. Tectonophysics 233: 163–176.

Pătrașcu, Ș., Șeclăman, M., and Panaiotu, C. 1993. Tectonic implications of paleomagnetism in Upper Cretaceous deposits in the Hațeg and Rusca Montană basins (South Carpathians, Romania). Cretaceous Research 14: 255–264.

Patrulius, D., Marinescu, F., and Baltreș, A. 1982. Dinosauriens ornithopodes dans les bauxites Néocomiennes de l'Unité de Bihor (Monts Apuseni). Anuarul Institutului de Geologie și Geofizică 59: 109–117.

Pedley, T. J. (ed.). 1977. *Scale Effects in Animal Locomotion.* Academic Press, New York. 545 pp.

Peláez-Campomanes, P., Damms, R., López-Martínez, N., and Álvarez-Sierra, M. A. 2000. The earliest mammal of the European Paleocene: the multituberculate *Hainina.* Journal of Paleontology 74: 701–711.

Pereda-Suberbiola, X. 1992. Armoured dinosaurs from the Late Cretaceous of southern France: a review. Revue de Paléobiologie, Vol. Spécial 7: 163–172.

Pereda-Suberbiola, X., Canudo, J. I., Cruzado-Caballero, P., Barco, J. L., López-Martínez, N., Oms, O., and Ruiz-Omeñaca, J. I. 2009. The last hadrosaurid dinosaurs of Europe: a new lambeosaurine from the Uppermost Cretaceous of Aren (Huesca, Spain). Comptes Rendus Palevol 8: 559–572.

Pereda-Suberbiola, X., and Galton, P. M. 1992. On the taxonomic status of the dinosaur *Struthiosaurus austriacus* Bunzel from the Late Cretaceous of Austria. Comptes Rendus de l'Académie des Sciences de Paris, Série 2, 315: 1275–1280.

Pereda-Suberbiola, X., and Galton, P. M. 1994. A revision of the cranial features of the dinosaur *Struthiosaurus austriacus* Bunzel (Ornithischia: Ankylosauria) from the Late Cretaceous of Europe. Neues Jahrbuch für Geologie und Paläontologie, Abhandlungen 191: 173–200.

Pereda-Suberbiola, X., and Galton, P. M. 1997. Armored dinosaurs from the Late Cretaceous of Transylvania. Sargetia 17: 203–217.

Persson, P. O. 1959. Reptiles from the Senonian (Upper Cretaceous) of Scania (south Sweden). Arkiv för Mineralogi och Geologi 2 (35): 431–478.

Peters, R. H. 1983. *The Ecological Implications of Body Size.* Cambridge University Press, Cambridge. 329 pp.

Petrescu, I., and Dușa, A. 1980. Flora din Cretacicul superior de la Rusca Montană—o raritate în patrimoniul paleobotanic național. Ocrotirea Naturii și a Mediului Înconjurător 24: 147–155.

Petrescu, I., and Duşa, A. 1982. Paleoflora din Senonianul Bazinului Rusca Montană. Dări de Seamă ale Şedinţelor, 3, Paleontologie 69: 107–124.

Petrescu, I., and Huică, I. 1972. Consideraţii preliminare asupra florei Cretacice de la Săsciori—Sebeş. Studii şi Cercetări de Geologie, Geofizică, Geografie, Serie Geologie 17: 461–464.

Pianka, E. R. 1970. On "*r*" and "*K*" selection. American Naturalist 104: 592–597.

Pianka, E. R. 1972. *r* and *K* selection or *b* and *d* selection? American Naturalist 106: 581–588.

Pincemaille, M. 1997. Un ornithopode du Crétacé supérieur de Vitrolles (Bouches-du-Rhône), *Rhabdodon priscus*. Diplôme d'Études Approfondies (DEA), Montpellier, France. 42 pp.

Plot, R. 1676. *The Natural History of Oxfordshire, Being an Essay toward the Natural History of England*. Published by the author, Oxford. 358 pp. (2nd edition, C. Brome, London, 1705).

Pol, C., Buscalioni, A. D., Carballeira, J., Francés, V., López-Martínez, N., Marandat, B., Moratalla, J. J., Sanz, J. L., Sigé, B., and Villatte, J. 1992. Reptiles and mammals from the Late Cretaceous new locality Quintanilla del Coco (Burgos Province, Spain). Neues Jahrbuch für Geologie und Paläontologie, Abhandlungen 184: 279–314.

Pol, D., Turner, A. H., and Norell, M. A. 2009. Morphology of the Late Cretaceous crocodylomorph Shamosuchus djadochtaensis and a discussion of Neosuchian phylogeny as related to the origin of Eusuchia. Bulletin of the American Museum of Natural History 324: 1–103.

Pop, G., Neagu, T., and Szasz, L. 1971. Senonianul din regiunea Haţegului (Carpaţii Meridionali). Dări de Seamă al Şedinţelor, 4, Stratigrafie 59: 115–169.

Pouyaud, L., Wirjoatmodjo, S., Rachmatika, I., Tjakrawidjaja, A., Hadiaty, R., and Hadie, W. 1999. A new species of coelacanth. Comptes Rendus de l'Académie des Sciences, Série 3, 322: 261–267.

Powell, J. E. 1992. Osteologia de *Saltosaurus loricatus* (Sauropoda—Titanosauridae) del noroeste Argentino. In: Sanz, J. L., and Buscalioni, A. D. (eds.). *Los Dinosaurios y su Entorno Biotico*. Instituto "Juan de Valdes," Cuenca, Spain. Pp. 165–230.

Prieto-Márquez, A., Gaete, R., and Galobart, A. 2000. A *Richardoestesia*-like theropod associated to a nesting site from the Maastrichtian of the Tremp Basin (northeastern Spain, pre-Pyrenees). Eclogae Geologicae Helvetiae 93: 497–501.

Prieto-Márquez, A., and Wagner, J. R. 2009. *Pararhabdodon isonensis* and *Tsintaosaurus spinorhinus*: a new clade of lambeosaurine hadrosaurids from Eurasia. Cretaceous Research 30: 1238–1246.

Prieto-Márquez, A., Weishampel, D. B., and Horner, J. R. 2006. The dinosaur

Hadrosaurus foulkii, from the Campanian of the East Coast of North America, with a reevaluation of the genus. *Acta Palaeontologica Polonica* 51: 77–98.

Rădulescu, C., and Samson, P.-M. 1986. Précisions sur les affinités des multituberculés (Mammalia) du Crétacé supérieur de Roumanie. Comptes Rendus de l'Académie des Sciences de Paris, Série 2, 303: 1825–1830.

Rădulescu, C., and Samson, P.-M. 1996. The first multituberculate skull from the Late Cretaceous (Maastrichtian) of Europe (Haţeg Basin, Romania). Anuarul Institutului Geologic al Romaniei (Suppl. 1: Abstracts of the 90th Anniversary Conference of the Geological Institute of Romania) 69: 177–178.

Rădulescu, C., and Samson, P.-M. 1997. Late Cretaceous Multituberculata from the Haţeg Basin (Romania). Sargetia 17: 247–255.

Rauhut, O. W. M., and Zinke, J. 1995. A description of the Barremian dinosaur fauna from Uña with a comparison to that of Las Hoyas. In: *II International Symposium on Lithographic Limestones, Cuenca, Spain: Extended Abstracts*. Ediciones de la Universidad Autónoma de Madrid, Spain. Pp. 123–127.

Raup, D. M., and Marshall, L. G. 1980. Variation between groups in evolutionary rates: a statistical test of significance. Paleobiology 6: 9–23.

Redelstorff, R., Csiki, Z., and Grigorescu, D. 2009. Nopcsa's heritage: dwarf status of Haţeg ornithopods supported by the histology of long bones. Journal of Vertebrate Paleontology 29 (Suppl.): 170A.

Reif, W.-E. 1980. Paleobiology today and fifty years ago: a review of two journals. Neues Jahrbuch für Geologie und Paläontologie, Monatshefte 1980: 361–372.

Rensch, B. 1959. *Evolution above the Species Level*. Columbia University Press, New York. 419 pp.

Riabinin, A. N. 1945. [Dinosaurian remains from the Upper Cretaceous of the Crimea.] Vsesoyuznyi Nauchno-Issledovatel'skii Geologicheskii Institut, Materialy po Paleontologii i Stratigrafii 4: 4–10. [In Russian, with an English summary.]

Ridley, M. 1994. *The Red Queen: Sex and the Evolution of Human Nature*. Penguin, New York. 404 pp.

Robel, G. 1966. *Franz Baron Nopcsa und Albanien*. Harrassowitz, Wiesbaden, Germany. 191 pp.

Robinet, J.-B. 1768. *Considérations philosophiques de la gradation naturelle des formes de l'être, ou les essais de la Nature qui apprend à faire l'Homme*. Saillant, Paris. 260 pp.

Ross, C. A., and Garnett, S. (eds.). 1989. *Crocodiles and Alligators*. Facts On File, New York. 240 pp.

Roth, V. L. 1992. Inferences from allometry and fossils: dwarfing of elephants on islands. Oxford Survey of Evolutionary Biology 8: 259–288.

Rudwick, M. J. S. 1985. *The Meaning of Fossils*. University of Chicago Press, Chicago. 287 pp.

Runcorn, S. K. 1963. Paleomagnetic methods of investigating polar wandering and continental drift. In: Munyan, W. H. (ed.). *Polar Wandering and Continental Drift*. Society of Economic Paleontologists and Mineralogists Special Publication 10. Pp. 47–54.

Rüpke, N. A. 1994. *Richard Owen: Victorian Naturalist*. Yale University Press, New Haven, CT. 462 pp.

Sachs, S., and Hornung, J. J. 2006. Juvenile ornithopod (Dinosauria: Rhabdodontidae) remains from the Upper Cretaceous (lower Campanian, Gosau Group) of Mutmannsdorf (Lower Austria). Geobios 39: 415–442.

Salgado, L., Coria, R. A., and Calvo, J. O. 1997. Evolution of titanosaurid Sauropods, I: phylogenetic analysis based on postcranial evidence. Ameghiniana 34: 3–32.

Sander, P. M. 1999. Life history of the Tendaguru sauropods as inferred from long bone histology. Mitteilungen aus dem Museum für Naturkunde der Humboldt-Universität Berlin, Geowissenschaftliche Reihe 2: 103–112.

Sander, P. M. 2000. Longbone histology of the Tendaguru sauropods: implications for growth and biology. Paleobiology 26: 466–488.

Sander, P. M., and Clauss, M. 2008. Sauropod dinosaur gigantism. Science 322: 200–201.

Sander, P. M., Klein, N., Buffetaut, E., Cuny, G., Suteethorn, V., and Le Loeuff, J. 2004. Adaptive radiation in sauropod dinosaurs: bone histology indicates rapid evolution of giant body size through acceleration. Organisms, Diversity & Evolution 4: 165–173.

Sander, P. M., Mateus, O., Laven, T., and Knötschke, N. 2006. Bone histology indicates insular dwarfism in a new Late Jurassic sauropod dinosaur. Nature 441: 739–741.

Sander, P. M., Peitz, C., Gallemi, J., and Cousin, R. 1998. Dinosaurs nesting on a red beach? Comptes Rendus de l'Académie des Sciences de Paris, Série 2, 327: 67–74.

Sanders, C. 1998. Tectonics and erosion, competitive forces in a compressive orogen: a fission track study of the Romanian Carpathians. PhD diss., Vrije Universiteits, Amsterdam. 204 pp.

Săndulescu, M. 1975. Essai de synthèse structurale des Carpathes. Bulletin de la Société Géologique de France 17: 299–358.

Sanz, J. L., Moratalla, J. J., Diaz-Molina, M., López-Martínez, N., Kälin, O., and Vianey-Llaud, M. 1995. Dinosaur nests at the sea shore. Nature 376: 731–732.

Sanz, J. L., Powell, J. E., Le Loeuff, J., Martinez, R., and Pereda-Suberbiola, X. 1999. Sauropod remains from the Upper Cretaceous of Laño (northcentral

Spain): Titanosaur phylogenetic relationships. Estudios del Museo de Ciencias Naturales de Álava 14: 235–255.

Schaeffer, B. 1952. Rates of evolution in the coelacanth and dipnoan fishes. Evolution 6: 101–111.

Schaffner, K. F. 1995. Comments on Beatty. In: Wolters, G., and Lennox, J. G. (eds.). *Concepts, Theories, and Rationality in the Biological Sciences*. University of Pittsburgh Press, Pittsburgh. Pp. 99–106.

Schindewolf, O. H. 1993. *Basic Questions in Paleontology* (translated by Schaefer, J.). University of Chicago Press, Chicago. 467 pp.

Schlüter, T. 1983. A fossiliferous resin from the Cenomanian of the Paris and Aquitanian basins of northwestern France. Cretaceous Research 4: 265–269.

Schmid, S. M., Berza, T., Diaconescu, V., Froitzheim, N., and Fügenschuh, B. 1998. Orogen-parallel extension in the Southern Carpathians. Tectonophysics 297: 209–228.

Schmidt-Nielsen, K. 1972. *How Animals Work*. Cambridge University Press, London. 114 pp.

Schmidt-Nielsen, K. 1975. Scaling in biology: the consequences of size. Journal of Experimental Zoology 194: 287–308.

Schmidt-Nielsen, K. 1979. *Animal Physiology: Adaptations and Environments*. Cambridge University Press, London. 560 pp.

Schopf, T. J. M. 1984. Rates of evolution and the notion of living fossils. Annual Review of Ecology and Systematics 12: 245–292.

Seeley, H. G. 1869. *Index to the Fossil Remains of Aves, Ornithosauria, and Reptilia, from the Secondary System of Strata Arranged in the Woodwardian Museum of the University of Cambridge*. Deighton, Bell, Cambridge. 143 pp.

Seeley, H. G. 1881. The reptile fauna of the Gosau Formation preserved in the Geological Museum of the University of Vienna. Quarterly Journal of the Geological Society of London 37: 620–702.

Seeley, H. G. 1883. On the dinosaurs from the Maastricht beds. Quarterly Journal of the Geological Society of London 39: 246–252.

Seeley, H. G. 1887a. On a sacrum apparently indicating a new type of bird, *Ornithodesmus cluniculus* Seeley. Quarterly Journal of the Geological Society of London 43: 206–211.

Seeley, H. G. 1887b. On the classification of the fossil animals commonly called Dinosauria. Proceedings of the Royal Society of London 43: 165–171.

Seeley, H. G. 1901. *Dragons of the Air: An Account of Extinct Flying Reptiles*. Methuen, London. 239 pp.

Sereno, P. C. 1991. *Lesothosaurus*, "fabrosaurids," and the early evolution of Ornithischia. Journal of Vertebrate Paleontology 11: 168–197.

Sereno, P. C. 1999. The evolution of dinosaurs. Science 284: 2137–2147.

Sereno, P. C., Beck, A. L., Dutheil, D. B., Larsson, H. C. E., Lyon, G. H., Moussa, B., Sadleir, R. W., Sidor, C. A., Varricchio, D. J., Wilson, G. P., and Wilson, J. A. 1999. Cretaceous sauropods from the Sahara and the uneven rate of skeletal evolution among dinosaurs. Science 186: 1342–1347.

Sereno, P. C., Dutheil, D. B., Iarochene, M., Larsson, C. E., Lyon, G. H., Magwene, P. M., Sidor, C. A., Varricchio, D. J., and Wilson, J. A. 1996. Predatory dinosaurs from the Sahara and Late Cretaceous faunal differentiation. Science 272: 986–991.

Shea, B. T. 1989. Heterochrony in human evolution: the case for neoteny reconsidered. Yearbook of Physical Anthropology 32: 69–101.

Shubin, N., and Alberch, P. 1986. A morphogenetic approach to the origin and basic organization of the tetrapod limb. Evolutionary Biology 20: 319–387.

Siddall, M. E. 1998. Stratigraphic fit to phylogenies: a proposed solution. Cladistics 14: 201–208.

Sigé, B., Buscalioni, A. D., Duffaud, S., Gayyet, M., Orth, B., Rage, J.-C., and Sanz, J. L. 1997. État des données sur le gisement Crétacé supérieur continental de Champ-Garimond (Gard, sud de la France). Münchener Geowissenschaften, Abhandlung A34: 111–130.

Simionescu, J. 1912. *Megalosaurus* aus der Unterkreide der Dobrogea (Rumänien). Centralblatt für Mineralogie, Geologie, und Palaeontologie 1913: 686–687.

Simpson, G. G. 1944. *Tempo and Mode in Evolution*. Columbia University Press, New York. 237 pp.

Simpson, G. G. 1953. *The Major Features of Evolution*. Columbia University Press, New York. 331 pp.

Smith, A. G., Hurley, A. M., and Briden, J. C. 1981. *Phanerozoic Paleocontinental World Maps*. Cambridge University Press, Cambridge. 102 pp.

Smith, A. G., Smith, D. G., and Funnell, B. M. 1994. *Atlas of Mesozoic and Cenozoic Coastlines*. Cambridge University Press, Cambridge. 99 pp.

Smith, J. L. B. 1956. *Old Fourlegs: The Story of the Coelacanth*. Longmans, Green, New York. 260 pp.

Smith, T., Codrea, V. A., Săsăran, E., Van Itterbeeck, J., Bultynck, P., Csiki, Z., Dica, P., Fărcaş, C., Folie, A., Garcia, G., and Godefroit, P. 2002. A new exceptional vertebrate site from the Late Cretaceous of the Haţeg Basin (Romania). Studia Universitatis Babeş-Bolyai, Geologia, Special Issue 1: 321–330.

Sober, E. 1991. *Reconstructing the Past: Parsimony, Evolution, and Inference*. Bradford, New York. 265 pp.

Sondaar, P. Y. 1977. Insularity and its effect on mammal evolution. In: Hecht, M. K., Goody, P. C., and Hecht, B. M. (eds.). *Major Patterns in Vertebrate Evolution*. Plenum, New York. Pp. 671–707.

Southwood, T. R. E., May, R. M., Hassell, M. P., and Conway, G. R. 1974. Ecological strategies and population parameters. American Naturalist 108: 791–804.

Stampfli, G. M. 1996. The Intra-Alpine terrain: a Paleotethyan remnant in the Alpine Variscides. Eclogae Geologicae Helvetiae 89: 13–42.

Stampfli, G. M., and Borel, G. D. 2002. A plate tectonic model for the Paleozoic and Mesozoic constrained by dynamic plate boundaries and restored synthetic oceanic isochrons. Earth and Planetary Science Letters 196: 17–33.

Stampfli, G. M., Borel, G. D., Cavazza, W., Mosar, J., and Ziegler, P. A. (eds.). 2001. *The Paleotectonic Atlas of the PeriTethyan Domain*. European Geophysical Society. Software available from Copernicus Publications, http://publications.copernicus.org/other_publications/cds_and_dvds.html.

Stanley, S. M. 1973. An explanation for Cope's Rule. Evolution 27: 1–26.

Starck, D., and Kummer, B. 1962. Zur Ontogenese des Schimpansenschädels (mit Bemerkungen zur Fetalisationshypothese). Anthropologische Anzeiger 25: 204–215.

Stearns, S. C. 1976. Life-history tactics: a review of the data. Quarterly Review of Biology 51: 227–243.

Stearns, S. C. 1983. The influence of size and phylogeny on patterns of covariation among life-history traits in the mammals. Oikos 41: 173–187.

Stearns, S. C. 1992. *The Evolution of Life Histories*. Oxford University Press, Oxford. 249 pp.

Stein, K., Csiki, Z., Curry Rogers, K. A., Weishampel, D. B., Redelstorff, R., Carballido, J. F., and Sander, P. M. 2010. Small body size and extreme cortical bone remodeling indicate phyletic dwarfism in Magyarosaurus dacus (Sauropoda: Titanosauria). PNAS: Proceedings of the National Academy of Sciences (USA) 107: 9258–9263.

Stevens, K. A., and Parrish, J. M. 1999. Neck posture and the feeding habits of two Jurassic sauropod dinosaurs. Science 284: 798–800.

Stiassny, M. L. J., Parenti, L. R., and Johnson, G. D. 1996. (eds.). *Interrelationships of Fishes*. Academic Press, San Diego. 498 pp.

Stilla, A. 1985. Géologie de la région de Haţeg-Cioclovina-Pui-Băniţa (Carpathes méridionales). Anuarul Institutului de Geologie şi Geofizică 66: 92–179.

Sues, H.-D. 1978. A new small theropod dinosaur from the Judith River Formation (Campanian) of Alberta. Zoological Journal of the Linnean Society 62: 381–400.

Suess, E. 1883–1901. *Das Antlitz der Erde*. 3 vols. F. Tempsky, Prague.

Suess, E. 1916. *Erinnerungen*. S. Hirzel, Leipzig, Germany. 451 pp.

Sulimski, A. 1968. Remains of Upper Cretaceous Mosasauridae (Reptilia) of central Poland. Acta Palaeontologica Polonica 13: 243–254.

Tallódi Posmoşanu, E., and Popa, E. 1997. Notes on a camptosaurid dinosaur from the Lower Cretaceous bauxite, Cornet—Romania. *Nymphaea* 23–25: 35–44.

Taquet, P. 1976. *Géologie et Paléontologie du Gisement de Gadoufaoua (Aptien du Niger)*. Cahiers de Paléontologie. Éditions du Centre National de la Recherche Scientifique, Paris. 191 pp.

Tasnádi-Kubacska, A. 1945. *Franz Baron Nopcsa*. Verlag des Ungarischen Naturwissenschaftlichen Museums, Budapest. 295 pp.

Taylor, M. P., Wedel, M. J., and Naish, D. 2009. Head and neck posture in sauropod dinosaurs inferred from extant animals. Acta Palaeontologica Polonica 54: 213–220.

Therrien, F. 2004. Paleoenvironmental reconstruction of the latest Cretaceous dinosaur-bearing formations of Romania. PhD diss., Johns Hopkins University. 309 pp.

Therrien, F. 2005. Palaeoenvironments of the latest Cretaceous (Maastrichtian) dinosaurs of Romania: insights from fluvial deposits and paleosols of the Transylvanian and Haţeg basins. *Palaeogeography, Palaeoclimatology, Palaeoecology* 218: 15–56.

Therrien, F. 2006. Depositional environments and fluvial system changes in the dinosaur-bearing Sânpetru Formation (Late Cretaceous, Southern Carpathians, Romania): post-orogenic sedimentation in an active extensional basin. *Sedimentary Geology* 192: 183–205.

Thomson, K. S. 1991. *Living Fossil: The Story of the Coelacanth*. W. W. Norton, New York. 252 pp.

Thulborn, R. A. 1982. Speeds and gaits of dinosaurs. Palaeogeography, Palaeoclimatology, Palaeoecology 38: 227–256.

Tomkins, C. 1996. *Duchamp: A Biography*. Henry Holt, New York. 444 pp.

Torrens, H. 1997. Politics and paleontology: Richard Owen and the invention of dinosaurs. In: Farlow, J. O., and Brett-Surman, M. K. (eds.). *The Complete Dinosaur*. Indiana University Press, Bloomington. Pp. 175–190.

Tsutakawa, R. K., and Hewett, J. E. 1977. Quick test for comparing two populations with bivariate data. Biometrics 33: 215–219.

Turner, A. 1997. *The Big Cats and Their Fossil Relatives*. Columbia University Press, New York. 234 pp.

Tykoski, R. S., and Rowe, T. 2004. Ceratosauria. In: Weishampel, D. B., Dodson, P., and Osmólska, H. (eds.). *The Dinosauria*, 2nd edition. University of California Press, Berkeley. Pp. 47–70.

Tzara, T. 1992. *Seven Dada Manifestos and Lampisteries* (translated by Wright, B.). Calder, New York. 118 pp.

Upchurch, P. 1995. The evolutionary history of sauropod dinosaurs. Philosophical Transactions of the Royal Society of London B, Biological Sciences 349: 365–390.

Upchurch, P. 1998. The phylogenetic relationships of sauropod dinosaurs. Zoological Journal of the Linnean Society 124: 43–103.

Upchurch, P., Barrett, P. M., and Dodson, P. 2004. Sauropoda. In: Weishampel, D. B., Dodson, P., and Osmólska, H. (eds.). *The Dinosauria*, 2nd edition. University of California Press, Berkeley. Pp. 259–322.

Valverde, J. A. 1964. Remarques sur la structure et l'évolution des communautés de vertébrès terrestres. La Terre et la Vie 111: 121–154.

Van Itterbeeck, J., Markevich, V. A., and Codrea, V. 2005. Palynostratigraphy of the Maastrichtian dinosaur- and mammal-bearing sites of the Râul Mare and Bărbat valleys (Haţeg Basin, Romania). Geologica Carpathica 56: 137–147.

Van Itterbeeck, J., Săsăran, E., Codrea, V., Săsăran, L., and Bultynck, P. 2004. Sedimentology of the Upper Cretaceous mammal- and dinosaur-bearing sites along the Râul Mare and Bărbat rivers, Haţeg Basin, Romania. Cretaceous Research 25: 517–530.

Van Valen, L. 1973a. A new evolutionary law. Evolutionary Theory 1: 1–30.

Van Valen, L. 1973b. Pattern and the balance of nature. Evolutionary Theory 1: 31–49.

Varricchio, D. J. 1997. Growth and embryology. In: Currie, P. J., and Padian, K. (eds.). *Encyclopedia of Dinosaurs*. Academic Press, New York. Pp. 282–288.

Vasse, D. 1995. *Ischyrochampsa meridionalis* n. g. n. sp., un crocodilien d'affinité gondwanienne dans le Crétacé supérieur du sud de la France. Neues Jahrbuch für Geologie und Paläontologie, Monatshefte 1995: 501–512.

Venczel, M., and Csiki, Z. 2003. New frogs from the latest Cretaceous of Haţeg Basin, Romania. Acta Palaeontologia Polonica 48: 609–616.

Venczel, M., and Gardner, J. D. 2005. The geologically youngest albanerpetontid amphibian, from the lower Pliocene of Hungary. Palaeontology 48: 1273–1300.

Vianey-Liaud, M. 1979. Les mammifères montiens de Hainin (Paléocène moyen de Belgique), part I: multituberculés. Paleovertebrata 9: 117–131.

Vianey-Liaud, M. 1986. Les multituberculés thanétiens de France, et leurs rapports avec les multituberculés nord-américains. Palaeontographica, Abteilung A, 191: 85–171.

Viany-Liaud, M., Mallan, P., Buscail, O., and Montgelard, C. 1994. Review of French dinosaur eggshells: morphology, structure, mineral and organic composition. In: Carpenter, K., Hirsch, K. F., and Horner, J. R. (eds.). *Dinosaur Eggs and Babies*. Cambridge University Press, New York. Pp. 151–183.

Vickaryous, M. K., Maryańska, T., and Weishampel, D. B. 2004. Ankylosauria.

In: Weishampel, D. B., Dodson, P., and Osmólska, H. (eds.). *The Dinosauria*, 2nd edition. University of California Press, Berkeley. Pp. 363–392.

Vidal, N., and Hedge, S. B. 2005. The phylogeny of squamate reptiles (lizards, snakes, and amphisbaenians) inferred from nine nuclear protein-coding genes. Comptes Rendus Biologies 328: 1000–1008.

Vrba, E. S., and Gould, S. J. 1986. The hierarchical expansion of sorting and selection: sorting and selection cannot be equated. Paleobiology 12: 217–228.

Vremir, M., and Codrea, V. 2002. The first Late Cretaceous (Maastrichtian) dinosaur footprints from Transylvania (Romania). Studia Universitatis Babeş-Bolyai, Geologia 2: 27–36.

Wayne, R. K. 1986. Cranial morphology of domestic and wild canids: the influence of development on morphological change. Evolution 40: 243–261.

Wegner, A. 1912. Die Entstehung der Kontinente. Geologische Rundschau 3: 276–292.

Weinberg, S. 2000. *A Fish Caught in Time: The Search for the Coelacanth*. HarperCollins, New York. 239 pp.

Weishampel, D. B. 1984. Evolution of jaw mechanisms in ornithopod dinosaurs. Advances in Anatomy, Embryology, and Cell Biology 87: 1–110.

Weishampel, D. B. 1985. An approach to jaw mechanics and diversity: the case of ornithopod dinosaurs. In: Duncker, H. R., and Fleischer, G. (eds.). *Functional Morphology in Vertebrates: Proceedings of the 1st International Symposium on Vertebrate Morphology*. Forschritt der Zoologie 30. Gustav Fischer, Stuttgart, Germany. Pp. 261–263.

Weishampel, D. B. 1990. Dinosaurian distributions. In: Weishampel, D. B., Dodson, P., and Osmólska, H. (eds.). *The Dinosauria*. University of California Press, Berkeley. Pp. 63–140.

Weishampel, D. B. 1991. A theoretical morphologic approach to tooth replacement in lower vertebrates. In: Vogel, K., and Schmidt-Kittler, N. (eds.). *Constructional Morphology and Biomechanics: Concepts and Implications*. Springer-Verlag, Berlin. Pp. 295–310.

Weishampel, D. B. 1995. Fossils, function, and phylogeny. In: Thomason, J. (ed.). *Functional Morphology in Vertebrate Paleontology*. Cambridge University Press, New York. Pp. 34–54.

Weishampel, D. B. 1996. Fossils, phylogeny, and discovery: a cladistic study of the history of tree topologies and ghost lineage durations. Journal of Vertebrate Paleontology 16: 191–197.

Weishampel, D. B. 2004. Ornithischia. In: Weishampel, D. B., Dodson, P., and Osmólska, H. (eds.). *The Dinosauria*, 2nd edition. University of California, Berkeley, California. Pp. 323–324.

Weishampel, D. B., Barrett, P. M., Coria, R. A., Le Loeuff, J., Xu X., Zhao X., Sahni, A., Gomani, E. M. P., and Noto, C. R. 2004. Dinosaur distribution. In: Weishampel, D. B., Dodson, P., and Osmólska, H. (eds.). *The Dinosauria*, 2nd edition. University of California, Berkeley, California. Pp. 517–606.

Weishampel, D. B., Grigorescu, D., and Norman, D. B. 1991. The dinosaurs of Transylvania: island biogeography in the Late Cretaceous. National Geographic Research and Exploration 7: 68–87.

Weishampel, D. B., and Heinrich, R. E. 1992. Systematics of Hypsilophodontidae and basal Iguanodontia (Dinosauria: Ornithopoda). Historical Biology 6: 159–184.

Weishampel, D. B., and Horner, J. R. 1990. Hadrosauridae. In: Weishampel, D. B., Dodson, P., and Osmólska, H. (eds.). *The Dinosauria*. University of California Press, Berkeley. Pp. 534–561.

Weishampel, D. B., and Horner, J. R. 1994. Life history syndromes, heterochrony, and the evolution of Dinosauria. In: Carpenter, K., Hirsch, K., and Horner, J. R. (eds.). *Dinosaur Eggs and Babies*. Cambridge University Press, New York. Pp. 229–243.

Weishampel, D. B., and Jianu, C.-M. 1996. New theropod dinosaur material from the Haţeg Basin (Late Cretaceous, western Romania). Neues Jahrbuch für Geologie und Paläontologie, Abhandlungen 200: 387–404.

Weishampel, D. B., and Jianu, C.-M. 1997. The importance of phylogeny in paleobiogeographic analyses, with examples from the North American hadrosaurids and European titanosaurids. Sargetia 17: 261–278.

Weishampel, D. B., and Jianu, C.-M. Unpublished manuscript. Franz Baron Nopcsa: a man out of time. Foreword to Weishampel, D. B., and Kerscher, O. *Franz Baron Nopcsa* [English translation of Tasnádi-Kubacska, A. 1945. *Franz Baron Nopcsa*].

Weishampel, D. B., Jianu, C.-M., Csiki, Z., and Norman, D. B. 2003. Osteology and phylogeny of *Zalmoxes* (n. g.), an unusual euornithopod dinosaur from the latest Cretaceous of Romania. Journal of Systematic Palaeontology 1: 65–123.

Weishampel, D. W., Mulder, E. W. A., Dortangs, R. W., Jagt, J. W. M., Jianu, C.-M., Kuypers-Peeters, M. M. M., Peeters, H. H. G., and Schulp, A. S. 2000. Dinosaur remains from the type Maastrichtian: an update. Geologie en Mijnbouw 78: 357–365.

Weishampel, D. B., and Norman, D. B. 1987. Dinosaur–plant co-evolution in the Mesozoic. *Fourth Symposium on Mesozoic Terrestrial Ecosystems*. Tyrrell Museum of Palaeontology, Drumheller, Alberta, Canada. Pp. 228–233.

Weishampel, D. B., and Norman, D. B. 1989. Vertebrate herbivory in the Mesozoic: jaws, plants, and evolutionary metrics. In: Farlow, J. O. (ed.). *Paleobiology of the Dinosaurs*. Geological Society of America Special Paper 238. Geological Society of America, Boulder, CO. Pp. 87–100.

Weishampel, D. B., Norman, D. B., and Grigorescu, D. 1993. *Telmatosaurus transsylvanicus* from the Late Cretaceous of Romania: the most basal hadrosaurid. Palaeontology 36: 361–385.

Weishampel, D. B., and Reif, W.-E. 1984. The work of Franz Baron Nopcsa (1877–1933): dinosaurs, evolution, and theoretical tectonics. Jahrbuch der Geologischen Bundes-Anstalt Wien 127: 187–202.

Weishampel, D. B., and White, N. M. (eds.). 2003. *The Dinosaur Papers: 1676–1906*. Smithsonian Institution Press, Washington, DC. 524 pp.

Weishampel, D. B., and Young, L. 1996. *Dinosaurs of the East Coast*. Johns Hopkins University Press, Baltimore. 275 pp.

Wellnhofer, P. 1980. Flugsaurierreste aus der Gosau-Kreide von Muthmannsdorf (Niederösterreich)—ein Betrag zur Kiefermechanik der Pterosaurier. Mitteilungen der Bayerische Staatssammlung für Paläontologie und Historische Geologie 20: 95–112.

Wellnhofer, P. 1991. *The Illustrated Encyclopedia of Pterosaurs*. Salamander, London. 191 pp.

Wellnhofer, P. 1994. Ein Dinosaurier (Hadrosauridae) aus der Oberkreide (Maastricht, Helvetikum-Zone) des bayerischen Alpenvorlandes. Mitteilungen der Bayerische Staatssammlung für Paläontologie und Historische Geologie 34: 221–238.

Westoll, T. S. 1949. On the evolution of the Dipnoi. In: Jepsen, G. L., Mayr, E., and Simpson, G. G. (eds.). *Genetics, Paleontology, and Evolution*. Princeton University Press, Princeton, NJ. Pp. 121–184.

Wiley, E. O., Siegel-Causey, D., Brooks, D. R., and Funk, V. A. 1991. *The Compleat Cladist*. University of Kansas, Museum of Natural History Special Publication 19. University of Kansas, Lawrence. 158 pp.

Willingshofer, E. 2000. Extension in collisional orogenic belts: the Late Cretaceous evolution of the Alps and Carpathians. PhD diss., Vrije Universiteit, Amsterdam. 146 pp.

Willingshofer, E., Andriesson, P., Cloething, S., and Neubauer, F. 2001. Detrital fission track thermochronology of Upper Cretaceous syn-orogenic sediments in the South Carpathians (Romania): inferences on the tectonic evolution of a collisional hinterland. Basin Research 13: 379–395.

Wilson, J. A. 2002. Sauropod dinosaur phylogeny: critique and cladistic analysis. Zoological Journal of the Linnean Society 136: 217–276.

Wilson, J. A. 2005. Overview of sauropod phylogeny and evolution. In: Curry

Rogers, K. A., and Wilson, J. A. (eds.). *The Sauropods: Evolution and Paleobiology*. University of California Press, Berkeley. Pp. 15–49.

Wilson, J. A., and Sereno, P. C. 1998. Early evolution and higher-level phylogeny of sauropod dinosaurs. Memoir (Society of Vertebrate Paleontology) 5: 1–68.

Wilson, J. A., and Upchurch, P. 2003. A revision of *Titanosaurus* Lydekker (Dinosauria, Sauropoda), the first dinosaur genus with a "Gondwana" distribution. Journal of Systematic Palaeontology 1: 125–160.

Wing, S. L., and Tiffney, B. H. 1987. The reciprocal interaction of angiosperm evolution and tetrapod herbivory. Review of Palaeobotany and Palynology 50: 179–210.

Witmer, L. M. 1991. Perspectives on avian origins. In: Schultze, H. P., and Trueb, L. (eds.). *Origin of the Higher Groups of Tetrapods*. Cornell University Press, Ithaca, NY. Pp. 427–466.

Witmer, L. M. 1995. The Extant Phylogenetic Bracket and the importance of reconstructing soft tissues in fossils. In: Thomason, J. (ed.). *Functional Morphology in Vertebrate Paleontology*. Cambridge University Press, New York. Pp. 19–33.

Witmer, L. M. 2001. Nostril position in dinosaurs and other vertebrates and its significance for nasal function. Science 293: 850–853.

Xu X., Zhou Z, and Prum, R. O. 2001. Branched integumental structures in *Sinornithosaurus* and the origin of feathers. Nature 410: 200–204.

Xu X., Zhou Z., and Wang X. 2000. The smallest known non-avian theropod dinosaur. Nature 408: 705–708.

Xu X., Zhou Z., Wang X., Kuang X., Zhang F., and Du X. 2003. Four-winged dinosaurs from China. Nature 421: 335–340.

Yarkov, A. A., and Nessov, L. A. 2000. [New records of hesperornithiform birds in the Volgograd Region.] Russkii Ornitologicheskii Zhurnal Ekspress Vypusk 94: 3–12. [In Russian.]

You H.-L., and Dodson, P. 2004. Basal Ceratopsia. In: Weishampel, D. B., Dodson, P., and Osmólska, H. (eds.). *The Dinosauria*, 2nd edition. University of California Press, Berkeley. Pp. 478–493.

You H.-L., Ji Q., Li J.-L., and Li Y.-X. 2003. A new hadrosauroid dinosaur from the mid-Cretaceous of Liaoning, China. Acta Geologica Sinica (English edition) 77: 148–154.

You H.-L., Luo Z.-X., Shubin, N. H., Witmer, L. M., Tang Z.-L., and Tang F. 2003. The earliest-known duck-billed dinosaur from deposits of late Early Cretaceous age in northwest China and hadrosaur evolution. Cretaceous Research 24: 347–355.

Zarski, M., Jakubowski, G., and Gawor-Biedowa, E. 1998. [The first Polish find of the Lower Paleocene crocodile *Thoracosaurus* Leidy, 1852: geological and palaeontological description.] Kwartalnik Geologiczny 42: 141–160. [In Polish.]

Ziegler, A. M., Parrish, J. M., Yao J-.P., Gyllenhaal, E. D., Rowley, D. B., Totman Parrish, J., Nie, S.-Y., Bekker, A., and Hulver, M. L. 1993. Early Mesozoic phytogeography and climate. Philosophical Transactions of the Royal Society of London B, Biological Sciences 341: 297–305.

Ziegler, A. M., Scotese, C. R., and Barrett, S. F. 1983. Mesozoic and Cenozoic paleogeographic maps. In: Brosche, P., and Sündermann, J. (eds.). *Tidal Friction and the Earth's Rotation, 2*. Springer-Verlag, Berlin. Pp. 240–252.

Ziegler, P. A. 1988. *Evolution of the Arctic-North Atlantic and the Western Tethys*. American Association of Petroleum Geologists Memoir 43. American Association of Petroleum Geologists, Tulsa, Oklahoma. 198 pp.

Ziegler, P. A. 1990. *Geological Atlas of Western and Central Europe*. Shell Internationale Petroleum Maatschappij B.V., The Hague. 130 pp.

Zweigel, P., Ratschbacher, L., and Frisch, W. 1998. Kinematics of an arcuate fold-thrust belt: the southern Eastern Carpathians (Romania). Tectonophysics 297: 177–207.

INDEX

Page numbers in *italics* indicate illustrations.

Abel, Othenio, 10, 86, 87, 105
Abelisaurus, 110
acipenseriforms (chondrosteans), 82, 83
actinistian fish, 154–56
Actinopterygii, 82–83
Acynodon, 67, *68*, 70
Adriatic Sea, *96*
Aegean Sea, *96*
Africa, 93, 95, 105, 112, 117, 169, 170, 179
agamids, 73
Ajkaceratops kozmai, 109
Alamosaurus, 28, 169
Albanerpeton, *81*, 82
Albanerpetontidae, 80, 81, *81*
Albania, 11–14
Albertosaurus, 110
Albian, 109, 217n15, 218n14
Alectrosaurus, 110
alligator lizards, 73
Alligatoroidea, 71
alligators, 19
Allodaposuchus, 71, 179
Allodaposuchus precedens, 67, *69*, 69–71
allopatry, 165
Allosaurus, 32
Alps, *94*
Alp-Tethys oceanic basin, 95
Altirhinus, 52
Alvarezsauridae, 31
alvarezsaurids, 37
ammonites, 98, 117
Amniota, 60
Ampelosaurus, 28, 107, 178, 179, *181*
Ampelosaurus atacis, 110, *180*
amphibians, 80–82, 107–9
Amphisbaenia, 73

Anabisetia, 45, *158*, 160, 161, 169
anatomy, 21–22, 114
angiosperms, 102, 103, 188
Anguimorpha, 72, 73
Anhanguera, 64
Ankylopollexia, 45
Ankylosauria, 11, 25, 38–43, 141, 161
Ankylosauridae, 39, 110
Ankylosaurus, 39
Anura, 80
Anurognathus, 64
Apatosaurus, 23, 133
Apoda, 80
Aptian, 95
Apulia, 96–98, 105
Apuseni Mountains, 15, 88, *94*, 97, 101
Arabia, 93
Aralosaurus, 52
Archaeopterydactyloidea, 64
Archaeopteryx, 19, 30, 31
Archosauria, 18, 19
Argentina, 27, 117, 161
Argentinosaurus, 117
arthropods, 122
Asia, 27; comparisons with, 110, 169; conditions in, 170, 171; and Hadrosauridae, 171, 174; and migration, 182, 183, 189, 192; and Transylvanian insularity, 112, 167
Auca Mahuevo, 56–58
Aude Valley, 174
Australia, 169, 170
Australopithecus afarensis, 18, 19
Australopithecus africanus, 19
Austria, 43, 109
Austrochelyidae, 76

Austrosaurus, 169
Aves, 18, 33
Avetheropoda, 31
Avialae, 31, 37
axolotl *(Ambystoma mexicanum)*, 118, *119*, 120
Azhdarchidae, 64, 65, 67, 109, 141, 184
Azhdarcho, 64
Azhdarchoidea, 64

Bactrosaurus, 52, *157*, 171, 172
Bactrosaurus johnsoni, 52
Bagaceratops kozmai, 176
Bakonydraco galaczi, 109
Bakony Mountains, 176
Balaur bondoc, 37, 38
Balkan Mountains, *94*
Bărăbanț, 15, 46, 73
Barbatodon transsylvanicus, 77, 79, 81
bat, 123
Batrachia, 80
Becklesius hoffstetteri, 72
Becklesius nopcsai, 72
beetle, dermestic, 104
Belgium, 108
Belgorod, 108, 109
benettitaleans, 102
Betasuchus bredai, 108
Bicuspidon hatzegiensis, 72
biogeography, 169–71, 184
birds, 19, 30, 31, 36
Black Sea, 1, *96*, 108
Boccaccio, Giovanni, 113
body size, 117, 118; and colonization, 142, 143, 185, 191, 192; and crocodilians, 118, 141; and Euhadrosauria, 132; and fauna, 86–87, 106; and France, 137; and *Iguanodon*, 45; and Iguanodontia, 130; and *Magyarosaurus*, 187; and migration, 143–45, 187–89; reasons for, 87, 118, 141, 142; and selection, 185; and *Struthiosaurus*, 187; and *Telmatosaurus*, 128–32, 187, 189; and *Telmatosaurus transsylvanicus*, 49; and Theropoda, 145; and *Zalmoxes*, 46, 187. See also dwarfism; growth and development
Bohemian Massif, 102

Borod Basin, 32
bowfin fish, 150
Brachiosaurus, 28, 133
Brachylophosaurus, 50
Bradycneme, 36
Broili, Ferdinand, 10
Brookes, Richard, 114
Buckland, William, 114–15, 116
Buffetaut, Eric, 25, 65, 70, 175
Bulgaria, 109
Bunzel, Emmanuel, 39, 40, 61

caecilians, 80
Camarasaurus, 28
Campanian, 95, 97
Campanian–Maastrichtian interval, 95
Camptosaurus, 25, 43–45, 129, 159, 160, 187
Carcharodontosaurus, 30, 117
Carnosauria, 31
Carroll, Lewis, 194–95
Casichelydia, 76
Čáslav, 109
Caucasus, 93
Caudata, 80
Cenozoic era, 94, 153
centipedes, 122
Cephalopoda, 19
Ceratopsia, 25, 111, 124, 161
Ceratosauria, 31
Cerna River, 87
chameleons, 73
character change, 131, 156–58; rates of, 150, 151, 159–62, 171. *See also* evolution
character optimization, 126–28, 130, 131, 170, 216n70
Cherven Bryag, 109
chimpanzee, 119, 120, *121*
Choceň, 60
choristoderans, 108
cladistics, 16, 18, 152, 168, 170, 171
cladogenesis, 165
cladogram, 18, 20, 21, 126–28, 151, 152
clam, unionid, 104
clubmoss (Lycopodiaceae), 103, *103*
Cochirleni, 204n4
Codrea, Vlad, 15, 34, 77
coelacanth, 148, 149, 154

292   Index

Coelophysoidea, 31
Coelurosauria, 31, 37
coevolution, 196, 197
colonization, 162, 168, 186; and body size, 142, 143, 185, 191–92; and clutch size, 191–92; and continental configurations, 182; and diversity, 106; in Europe, 185; and founder effect, 166; and loss of genetic variation, 186; and sexual maturation, 186, 191; success in, 167, 183, 184, 186, 192
Comoros Islands, 148, 149, 162
Compsognathidae, 31
Compsognathus, 30, 31
conifers, 102, 104
continental drift, 89–91
Cope, Edward Drinker, 117, 142
Cope's Rule, 117, 142
Cornet, 204n4
Courtenay-Latimer, Marjorie, 147–48, 154
Cretaceous, 93; ephemeral terrestrial habitats of, 159; flora of, 102; habitats of, 105; and hadrosaurids, 52; and isolation and colonization of Transylvanian region, 145; mammals in, 77; and multituberculates, 77; and ornithopods, 45; and Pterosauria, 60, 64; and sauropods, 28; and Southern Carpathian Mountains, 94, 97; and titanosaurs, 27
Cretaceous-Tertiary boundary, 65, 77, 198
Crimea, 108
crocodilians, 3, 19, 98, 179, 180; and *Allodaposuchus precedens*, 68–70; body size of, 118, 141; in Europe, 107–9; eusuchian, 109; feeding by, 72; growth of, 184; social behavior of, 72
Crocodilus affulvensis, 68
Crocodylia, 18, 19, 71
Crocodyloidea, 71
Cryptodira, 76, 77
Csiki, Zoltan, 27, 34, 44, 65, 124
Ctenodactylidae, 64
cuckoo *(Cuculus canorus)*, 197
Cuvier, Georges, 114
cycads, 102
Czechia, 60, 102, 109, 211n57

Danube, 1
Daspletosaurus, 110
deer, 119
Deinonychus, 30, 36
Deinotherium giganteum, 114
Densuş, 15
Densuş-Ciula Formation, 37, 63, 67, 100–102, 159
Devonian, 93
Diacryptodira, 76
Dietz, Robert S., 91, 92
Dinornis maximus, 22
Dinosauria, 18, 19, 22–25, 116, 142
Diplodocoidea, 28, 110
Diplodocus, 117, 133
diversity, 16, 106, 109, 110, 165, 169
Djungaripteroidea, 64
dogs, 122
Dollo, Louis, 10, *10*
Dollodon, 52
donkeys, 119
Doratodon, 67, *68*, 70, 71
dortokids, 77
Drasković, Ludwig Graf, 11–12
Dromaeosauridae, 31, 33, 34, *36*, 36–38
Dromaeosaurus, 35
Dryomorpha, 45, *158*
Dryosaurus, 45, 159, 160, 187
Dsungaripteridae, 64
Dsungaripteroidea, 63, 64
Dsungaripterus, 64
dwarfism, 142–46, 184–87, 191, 192, 213n29; and endocrine system, 87, 118, 141, 150; and femoral structure, 124, 125; and insular isolation, 86, 87, 142, 143. *See also* body size; growth and development

Early Cretaceous, 169, 179; migrations in, 183, 184; and Neotethyan Ocean, 97; and Severin Basin, 95; and *Telmatosaurus*, 159
Ebro High, 97
Edmontonia, 39, 40
eggs, 107, 108, 190–92; angusticaniculate, 55; and clutch size and colonization, 191, 192; dendrospherulitic, 55; and eggshell construction, 54, 55, 58; hatchlings from, 55, 56; nests of, 53–58

Index   293

elephants, 114, 119, 143–45
Elopteryx, 36
Elopteryx nopcsai, 32–33, *34*, 36
embryos, 55–58
enantiornithine bird, 109
England, 98, 115, 179, 211n57
Eolambia, 52
Eolambia caroljonesa, 171
Euhadrosauria, 52, 131, 171–74, 189; and body size, 132; clutch size of, 192; ghost lineage of, *157*
Euoplocephalus, 39
euornithopod, 176
Euramerica, 179, 180
Euronychodon, 36, *36*, 37
Europe, 27, 93, 97, 174, 182; and dinosaur distribution, 170, 171, 181; and migration, 182, 185, 189; as source area, 169, 178, 179
Eusuchia, 70, 71, 109, 179
evolution, 16–18, 21, 120, 142, 150, 151, 154, 156. *See also* character change; morphological change, rates of

Fărcădin, 87
Fennosarmatia, 97
fern (Polypodiaceae), *103*
fern (pteridophytes), 102
fern, filicalean (Gleicheniaceae), 103
fern, filicalean (Polypodiaceae), 103
Fontllonga hadrosaurid, 174
foraminifera, 98
France, 137, 169, 174, 175, 182; and *Ampelosaurus*, 110, 178, 181; fauna of, 106–10, 175
Fritsch, Anton, 60
frogs, 80, 81

Gallodactylidae, 64
gar *(Lepisosteus oculatus)*, *83*
Gargoyleosaurus, 39
Gasparinisaura, 45, 169
Gastonia, 39
Gavialoidea, 71
Gavials, 19
Gekkota, 73
Germanodactylus, 64
ghost lineage, 151–54, 171
ghost lineage duration (GLD), 152, 153, 159, 160

Giganotosaurus, 30, *32*, 110, 117
gigantism, 86, 87, 117, 141, 142, 145
Gila monsters, 73
ginkgoes, 102
Gobisaurus, 39, 169
Gondwana, 27, 93, 171, 178
gorilla, 119, 120
Gosau Beds (Austria), 43
Gosau fauna, 61, 108
Gould, Stephen J., 120, 146, 163, 199
Grigorescu, Dan, 15, 33, 34, 65, 101, 124
growth and development: acceleration of, 122, 123; of crocodilians, 184; of Hadrosauridae, 126; of Hatzegopteryx, 184; among herbivores, 145; of *Hypsilophodon*, 137–40; lines of arrested (LAGs), 124; of *Magyarosaurus*, 126, 133–37, 141, 144, 146, 184; of *Magyarosaurus dacus*, 125; of Mammalia, 184; of ornithopods, 125; of *Orodromeus*, 137–40; of pterosaurs, 184; of *Rhabdodon*, 137–40; of sauropods, 133–37; and selection, 185; of *Struthiosaurus*, 184; of *Telmatosaurus*, 126, 128–32, 135, 137, 141, 144, 146, 184; of *Telmatosaurus transsylvanicus*, 124–25; of *Tenontosaurus*, 138–40; of Theropoda, 184; of titanosaurian sauropods, 126; of titanosaurs, 137; of turtles, 184; of *Tyrannosaurus rex*, 124; of *Zalmoxes*, 126, 137–40, 146; of *Zalmoxes robustus*, 125, 139–41, 184; of *Zalmoxes shqiperorum*, 138–40, 184. *See also* body size; dwarfism
Groza, Ioan, 15
Gryposaurus, 53
Gymnophiona, 80

Hadrosauridae: cranial adornments of, 53; dentition of, 45, *51*, 129–32, 187–90, 192; eggs of, 56; in Europe, 108, 109, 174; evolutionary relationships in, 52; Fontllonga, 174; growth of, 126; and Ornithischia, 23; and *Pararhabdodon*, 174; perinatal bones of, 58; reproductive biology of, 190–92; and Rhabdodontidae, 46; and social behavior, 53; source area of, 169, 171–

72, 174; in Spain, 176; and
Telmatosaurus, 132, 157; and
Telmatosaurus transsylvanicus, 49–51; and *Tethyshadros insularis*, 172
Hadrosaurinae, 50–52, 131
Hadrosauroidea, 52, 109
Haeckel, Ernst, 120
Haeckel's Biogenic Law, 120
Hainina, 77
Harrison, Colin J. O., 33
Hațeg: amphibians of, 81–82; ankylosaur material from, 40; dromaeosaurid from, 34, *35;* eggs from, 58; fauna of, 15; formations in, 99–101; fossils from, 2, 25, 32, 36; geography of, 98–105; geology of, *100;* history of, 6; and insular isolation, 84–88; as island, 105–11; mammals of, 77–81; mesoeucrocodilians of, 67–73; paleoecological context of, 16; perinatal bones known from, 58; and plate tectonics, 97; pterosaurs from, 59–67; *Rhabdodon priscus* from, 43; river systems of latest Cretaceous in, 101; squamates of, 72–77; *Struthiosaurus transylvanicus* from, 38; and *Telmatosaurus transsylvanicus*, 49; turtles from, 73–77; work in, 15; *Zalmoxes* from, 46
Hatzegobatrachus grigorescui, 81
Hatzegopteryx, 64, 66, 72, 141, 184
Hatzegopteryx thambema, 65–66
Hennig, Willi, 18, 151
Heptasteornis, 36
Heptasteornis andrewsi, 33, 36
Herrerasaurus, 24
hesperornithids, 109
heterochrony, 119–26, 134, 144, 146, 168
hippos, dwarfed, 119
historical contingency, 168, 190, 198, 199
Holden, John C., 91, 92
Hominidae, 18, 19
Homo erectus, 19
homology, 17–21
homoplasies, 19–21
Homo sapiens, 18, 19, 118
Horner, Jack, 58, 190
horsetails (sphenophytes), 102
Huene, Friedrich von, 11, *11,* 25, 27

humans, 119–21
Hunedoara County, 2
Hungarosaurus tormai, 109, 176
Hungary, 108, 109, 175, 176, 181
Hylaeochampsa, 179
Hylaeosaurus, 38, 169
Hypacrosaurus, 190
hypermorphosis, 122
Hypsilophodon, 44, 45, 137–40, 169
Hypsilophodon foxii, 139
Hypsilophodontidae, 44

Iberia, 97, 107, 182
Iguania, 73
iguanians, 73
iguanids, 73
Iguanodon: body size of, 45; and hadrosaurids, 131, 187; lower teeth of, 128, 129; and migration, 169; as ornithischian, 23; recognition of, 116; and *Rhabdodon,* 43, 44; and Rhabdodontidae, 46, 160; teeth of, 130; and *Telmatosaurus,* 49, 50
Iguanodon bernissartensis, 52
Iguanodontia: body size of, 130; and Hadrosauridae, 171, 187; and Ornithopoda, 45; rate of evolutionary change of, 160; and Rhabdodontidae, 46, 159, 176; teeth of, 130, *131;* and *Telmatosaurus transsylvanicus,* 157
Iguanodontoidea, 52
Iharkut, 176, 177
Iharkutosuchus makadi, 109
Indonesia, 149
insects, 104, 122
invertebrates, 104
Irish Massif, 98
Isisaurus, 28
island, evidence of, 105–11
Island Rule, 143–45
isolation: and clutch size, 191, 192; and dwarfism, 86, 87, 143–45, 185–88; effects of, 106; and Europe, 185; insular, 86, 87; and sexual maturity, 144, 146, 186; and speciation, 165–67; and *Telmatosaurus,* 162; and Transylvanian region, 111, 162, 167, 168, 192
Italy, 70, 106, 108, 172, 174, 182

Index 295

Jaekel, Otto, 10
Janenschia, 124
Jeyawati, 52, *157*, 171
Jibou, 15, 46, 73
Jibou Formation, 46, 101
Jinzhousaurus, 52
Jiu River, 87
Jurassic, 43, 114, 115, 124; and Albanerpetontidae, 80; and iguanodontians, 159; and *Kallokibotion*, 180; and Mammalia, 78; and Neotethyan Ocean, 88; and Paleotethyan Ocean, 93; and pterosaurs, 60, 64; and *Struthiosaurus*, 182; and Testudinata, 76; and *Zalmoxes* and *Rhabdodon*, 176, 182, 183

Kadić, Ottokar, 67, 68
Kallokibotion, 3, 11, 76, 179
Kallokibotion bajazidi, 74, *75*
Kayentachelys, 76
Kessler, Eugen, 33
K life-history strategy, 190–92
kogaionids, 179–81
Kogaionon ungureanui, 77, *79*, 81
Komodo lizards, 119
Kordos, László, 128
Kutná Hora, 109

Labirinta Cave, 109
Laevidoolithidae, 55
Lambeosaurinae, 50–52, 131
Lambrecht, Kalman, 33
Lancrăm, 15, 46, 73
Late Cretaceous, 16, 82, 161–62; and Carpathian Mountains, 98; duckbills from, 49; in Europe, 108–10; fauna of, 36; invertebrates from, 104; and *Kallokibotion*, 180; landscape of, 84; migrations in, 171; of Mongolia, Argentina, and the United States, 37; and Neotethyan Ocean, 97; of North America, 34; and Sânpetru Formation, 2, 3; and *Telmatosaurus*, 159; theropods from, 30, 32, 38; and titanosaurs, 27; of Transylvania, 32; vertebrates of, 10
Late Jurassic, 45, 117; and Apulia and Rhodope, *95, 96;* and multituberculates, 77;

and *Struthiosaurus*, 177; and *Telmatosaurus*, 183; and titanosaurs, 27
Late Triassic, 28, 76, 93
Latimeria, 154–56, 161, 162
Latimeria chalumnae, 148–50
Latimeria menadoensis, 149
Laurasia, 38, 93, 170
laurels (Lauraceae), *103*, 104
Leaellynasaura, 169
Le Loeuff, Jean, 25, 175
Lepisosteidae, 82
Levnesovia, 52, *157*, 171, 172
Levnesovia/Bactrosaurus clade, 160
Linnaeus, Carolus, 17
Lirainosaurus, 28, 107, 178, 179, *181*
Lirainosaurus astibiae, 176, *180*
Lissamphibia, 80, 82
living fossil, 150, 151, 154, 162
lizards, 72, 108, 109
Lower Quarry, 3

Maastrichtian, 84, 88, 104, 108, 109, 159, 173, 174, 180
Macronaria, 28
Macropoma, 154
Madagascar, 27
madtsoiids, 73
Magyarosaurus, 25, *26*, 27, *29*, 168, 183; anatomy of, 30; and body size, 187; and dwarfing, *136*, 184; eggs of, 56; and Europe, 178–79, *181*, 182; growth and development of, 126, 133–37, 141, 144, 146; and Nopcsa, 11; as predator, 37; rarity of, 59; rate of evolutionary change of, 161; and Sauropoda, 28
Magyarosaurus dacus, 16, 26, 27, 30, 110, 118, 125, 215n65
Magyarosaurus hungaricus, 27
Magyarosaurus transsylvanicus, 27
Maiasaura, 50, 55, *129*, 130, 190
Malawisaurus, 28, 133, 169, 179, 182
Mammalia, 77–81, 107, 141; in Cretaceous, 77; evolutionary relationships in, 78; growth of, 184; placental, 108
Maniraptora, 31, 36, 62
Maniraptoriformes, 31
Mantell, Gideon, 115–16
Mantellisaurus, 52
marsupials, 77, 78

Massif Central, 97
Matheron, Philippe, 43, 46, 68–69
Mayor, Adrienne, 113
Mayr, Ernst, 165–66
Megaloolithidae, 56
Megalosaurus, 16, 23, 31, 32, 115, 116
Megalosaurus cf. *superbus*, 204n4
Megalosaurus hungaricus, 32
Meiolaniidae, 76
Mesoeucrocodylia, 67–73
Mesozoic, 30, 81, 117; and *Apatosaurus*, 133; and fossil record, 22, 153; and multituberculates, 77; and Neotethyan Ocean, 93; and pterosaurs, 59; and Romania, 105
Mesozoic mammals, 77
Mezholezy, 108, 109
Middle Jurassic, 16; and Albanerpetontidae, 82; and Ornithopoda, 45
migration, 169–84; and Africa, 179, 182; and Asia, 182, 183, 189, 192; and body size, 143–45, 184–89; and Cretaceous, 171, 183, 184; and Europe, 177–78, 182, 185, 189, 198; and historical contingency, 198; and North America, 177–78, 182, 183, 189, 192; and South America, 179; and titanosaur sauropods, 30
millipede, 122
Minmi, 39, 169
Miocene, 80, 82, 96, 98
moas, 119
Mochlodon, 43, 44
Mochlodon suessi, 108
Moesia, 95, 97
Moesian microplate, platform, promontory, 94–96
mollusks, 104
monitors, 73
monophyly, 17–18, 152
Monotremata, 78
morphological change, rates of, 151, 154–56. *See also* evolution
mosasaurs, 98, 108, 210n36
mouse, 77
Multituberculata, 77, 78, 80–81, 180–81
Mureş River, 101
Muthmannsdorf, Austria, 60, 61, 108
Muttaburrasaurus, 169

Nălaţ Vad, 46
natural selection, 120, 166, 188, 195, 197–99
Naturphilosophie, 120
nautilus, pearly, 117
Nemegtosaurus, 28, 182
neoceratopsians, 109
neo-Lamarckian inheritance, 141, 150
Neopterygii, 82
Neosauropoda, 28
Neosuchia, 71
neoteny, 123
Neotethyan Ocean, 84, 85, 88, 93, 97, 98, 105, 143
Netherlands, 108
Neuquensaurus, 28
newts, 80
Nodosauridae, 38–43, 107–10, 175, 176
Nopcsa, Franz, *11*, 25, 32, 38, 40, 68, 70, 84, 156, 168, 181; and body size, 125, 128–29, 133, 141–42, 143, 150; career of, 8–14; and dwarfism and insular isolation, 85–88; and global positions of the continents, 89; and Haţeg area, 161; and Haţeg as island, 105; and pterosaurs, 62; and *Rhabdodon priscus*, 43–44; and source areas, 169, 171; and *Struthiosaurus*, 42; and *Telmatosaurus transsylvanicus*, 9–10, 49; and turtles, 74, 77; and Wegener, 90; and *Zalmoxes*, 43
Nopcsa, Ilona, 8
Norell, Mark, 151
Noripterus, 64
Norman, David, 44
North America, 110, 117, 167, 169–71; and Hadrosauridae, 171, 174; and migration, 177, 178, 182, 183, 189, 192; and Transylvanian insularity, 112
North Atlantic, 93, 97
North Sea, 97
Nyctosauridae, 64

Oarda de Jos, 15, 40, 46, 73, 74, 77
Ohaba Sibişel, 2
Oligocene, 77, 78, 88, 98, 103
ontogeny, 120, 122
Opisthocoelicaudia, 28, 169
opossums, 150

Oradea, 33
Ornithischia, 19, 23–25, 38
Ornithocheiridae, 62
Ornithocheiroidea, 64
Ornithocheirus, 60–62, 64
Ornithocheirus bunzeli, 61
Ornithocheirus hlavatschi, 60, 61
Ornithodesmus, 62
Ornithodesmus cluniculus, 207n7
Ornitholestes, 31
Ornithomimosauria, 31, 37
ornithomimosaurs, 110
Ornithopoda, 2, 107–9, 169; cladogram of, *172;* evolutionary relationships in, 45; growth of, 124, 125; iguanodontian, 137; and Nopcsa, 10–11, 16; and rate of evolutionary change, 160, 161; and southern France, 175; teeth of, 129, 130; Transylvanian, 43–53; and Transylvanian deposits, 25
Orodromeus, 45, 137–39, 140
Orodromeus makelai, *139*
Orthomerus dolloi, 108
Orthomerus weberi, 108
ostracodes, 104
Ouranosaurus, 50, 52, 129–31, 169
Oviraptorosauria, 31, 110
Owen, Richard, 21–22, 116
owls, 33, 36

pachycephalosaurs, 25, 110, 111
Pâclişa, 101
paedomorphosis, 118, 120, 122, 123, 125, 139, 144, 146. *See also* body size; growth and development
paleobiology, 9, 10
paleoclimatology, 93
Paleotethyan Ocean, 93
Paleozoic, 92, 93
palm (Arecaceae), *103*, 104
Paludititan nalatzensis, 27
Pangea, 92–93
Panoplosaurus, 39, 40
Paralatonia transylvanica, 81
Paranthropus robustus, 19
parapatry, 165
paraphyly, 18
Pararhabdodon, 107, 108, 175
Pararhabdodon isonensis, 110, 173, 174

Parasaurolophus, 53
Paraves, 31
Paris Basin, 97
Parksosaurus, 45, 160
Paronychodon, 36, 37
parrots, 31
parsimony, 21
Patagonia, 133
Pawpawsaurus, 39, 169
pelican, 33
Pelorosaurus, 169
peramorphosis, 117, 118, 120, 122, 123, 131, 139. *See also* body size; growth and development
Pereda-Suberbiola, Xabier, 25, 40
Peri-Tethyan realm, 106
Petroşani Basin, 98
phylogeny, 169; and evolutionary rates, 156, 160, 161; and geographic data, 170; hierarchy of, 17; and homology, 19; and homoplasies, 19; and ontogeny, 120, 122; single origin in, 17; size increases through, 142
Pinacosaurus, 39
Pincemaille, Marie, 46
placentals, 77, 78
Plateosaurus, 24, 124
plate tectonics, 89–98, 176, 183, 197; and Transylvania, 105–6, 111–12, 154, 167
Pleistocene, 113, 143, 145
plesiosaurs, 108
Pleurodira, 76, 77
Pliocene, 98
Plot, Robert, 114
Poiana-Руscă Mountains, 87, 88, 98, 99
Polish Trough, 97
Polypteridae, 82
Portugal, 107
predators, 30, 43, 145, 167, 185–87, 189
Probactrosaurus, 52, 169
proboscideans, 143
Proganochelys, 76
progenesis, 122, 123
prosauropods, 124
Protohadros, 52
Protohadros byrdi, 171
pseudoacromion, 40
Psittacosaurus, 124

Pteranodon, 63, 64
Pteranodontidae, 66, 67
Pteranodontoidea, 64
Pterodactyloidea, 60, 64
Pterodactylus, 64
Pterosauria, 19, 59–67, 107–9, 141; bones of, 65–66; crest of, 67; evolutionary relationships in, 64; feeding by, 67; femur of, 63; gait of, 66, 67; growth of, 184; notarium of, 62, *63;* as predator, 72; social behavior of, 67; warm-bloodedness of, 67; wings of, 65
Pui, 27, 77, 87, 100

Quaesitosaurus, 28
Quetzalcoatlus, 64
Quetzalcoatlus northropii, 66

Rapetosaurus, 28, 133, 179, 182
Rapetosaurus krausei, 27
Râul Bărbat, 100
Râul Mare, 100
red-backed shrike *(Lanius collurio)*, 197
Red Queen hypothesis, 169, 196–99
Retezat Mountains, 1, 87, 88, 98, 99, 101, 168
Rhabdodon, 10–11, 45, 107, 108, 175–77; ghost lineage of, *158*, 159; growth and development of, 137–40; migration of, 182; source area for, *178*
Rhabdodon priscus, 43, 44, 46, *139*, 160, 176
Rhabdodontidae, 45, 46, 108, 109, 159, 176, 177; and migration, 182; and rate of evolutionary change, 160
rhamphorhynchoids, 64
Rhaptochelydia, 76
Rhenish Massif, 98
Rhodope, 95–97
Rhodope microplate, *96*, 98
Rhodopian Mountains, *94*
Riabinin, A. N., 108
Richardoestesia, 36, 37
Rocasaurus, 28, *136*, 179
Romania, 1, 105, 110
Roth, V. Louise, 143, 144, 145, 146, 186
Roth model, 186
r selection, 186, 192
r strategy, 190, 191

Rusca Montană Basin, 98, 103
Russia, 108, 109, 174

Saichania, 39
salamander, 80
salamander *(Ambystoma mexicanum)*, *119*
Saltasauridae, 28
Saltasaurus, 28, 133, 169, 179
Saltasaurus loricatus, 27
Sânpetru, 1, 2, 60, 63, 77, 159
Sânpetru Formation, 2, 3, 6, 40, 99–102
Sântămărie Orlea, 100
Santonian, 32, 109
Șard Formation, 46, 101, 102
Saurischia, 19, 23
Saurolophus, 53
Sauropelta, 39
Sauropoda, 3, 11, 16, 28, 37, 110, 161; braincase of, 27, 29; embryonic remains of, 57; growth and development of, 117, 124, 133–37; and Saurischia, 23; in Spain, 176
Sauropodomorphs, 19
Saurornithoides, 36
Saurornitholestes langstoni, 34
Schindewolf, Otto H., 142
Scincomorpha, 72, 73
Scleroglossa, 73
screw pines (Pandanaceae), 104
Sebecosuchia, 71
Sebeș, 15, 46, 47, 73, 101
Sebeș Formation, 37
Sebeș Mountains, 87
Seeley, Harry Govier, 22–23, 39–40, 43, 60–61
Seismosaurus, 117
Selmacryptodira, 76
Serpentes, 73
Severin Ocean, 95
sexual maturity, 123, 125, 144, 146, 186, 191, 192
Shamosaurus, 39
Sibişel River, 1–3, 15, 33, 43, 87, 100
Sichuan (China), 133
Silvisaurus, 39
Simpson, G. G., 150
skinks, 73
Slovenia, 108

Smith, J. L. B., 148
snakes, 72, 73, 108
South America, 110, 133, 169, 170, 179
South Atlantic, 93
Southern Carpathian Mountains, 1, *94*, *96*, 97, 98
Spain, 107–9, 110, 137, 173–76, 181
speciation, 164–66
sphenophytes, 103
Spinosauroidea, 31
Squamata, 73
Stegosauria, 25, 38, 161
Steno, Nicholas, 16
Struthiosaurus, 11, 107, 108, 110, 141, 161, 168, 183; and digestion, 42; discovery of, 39, 40; growth of, 184, 187; and locomotion, 42, 43; and migration, 182; and North America, 177; rarity of, 59
Struthiosaurus austriacus, 40, 43, 110
Struthiosaurus languedocensis, 110
Struthiosaurus transylvanicus, 16, 38–43, 110
sturgeon *(Acipenser transmontanus)*, 83
Suess, Eduard, 8, 9, 87, 89
Sulawesi Island, 149, 162
Sum of Minimum Implied Gaps (SMIG), 152
Şureanu Mountains, 98
Sweden, 211n57
sycamore (Platanaceae), *103*, 104
sympatry, 165

Tapejara, 64
Tapejaridae, 64
Tapejaroidea, 64
Tarascosaurus, 107
Tarbosaurus, 110
taxonomic rates, 150, 151
Teleosteii, 82, 83, 98, 154
Telmatosaurus, 33, 52, 59, 109, *132*, 168, 169, 172, 192; and body size, 128–32, 187, 189; and dentition, *129*, 146, 190; and diet, 189; eggs of, 56; and Europe, 171; and *Fontllonga hadrosaurid*, 174; ghost lineage of, *157*, 159; growth and development of, 126, 135, 137, 141, 144, 146, 184; migration of, 182, 183; and *Pararhabdodon*, 174; and rate of evolutionary change, 159–62; scapulocoracoid of, 3
Telmatosaurus transsylvanicus, 9–10, *9*, 16, 49–53, 58, 110; ghost lineage of, 156–57; growth of, 124, 125; source area of, 171
Tendaguru (Tanzania), 133
Tenontosaurus, 44–46, 137, 160, 169, 187; ghost lineage of, *158;* growth and development of, 138–40; lower teeth of, 129; and rate of evolutionary change, 160, 161
Tenontosaurus tilletti, *139*
Tertiary, 77, 96, 97, 122
Testudinata, 76, 180
Testudines, 76, 77
Tetanurae, 31
Tethyan Ocean, 92, 93
Tethyan region, 93–98
Tethyshadros, 52, *132*, *157*, 160, 171, 172, 174
Tethyshadros insularis, 108, 172, 173
tetra, blue neon *(Paracheirodon simulans)*, 83
tetrapods, 111, 149
Theria, 78
Theriosuchus, 71
Theriosuchus sympiestodon, 67, *68*
Therizinosauroidea, 31, 110
Theropoda, 3, 16, 19, 107–9, 117; alvarezsaurids in, 37; avialan, 175; avian, 161; biogeographic dynamics of, 181; dromaeosaurid, 175; dromaeosaurid and troodontid, 141; in Europe, 110; evolutionary relationships in, 31; growth of, 124, 145, 184; nonavian, 161; and Ornithischia, 23; ostrich-mimicking, 31; and southern France, 175; in Spain, 176; Transylvanian, 30–38
Thescelosaurus, 45, 160, 169
thyreophorans, 161
tigers, Bali, 119
Titanosauria, 2, 25–30, 107, 109, 133; anatomy of, 27, 28; cladogram of, *136;* eggs of, 56; embryos of, 57, 58; and Europe, 182; growth and development of, 126, 137; source area for, 178; and southern France, 175

Titanosaurus indicus, 27
titanotheres, 122
toads, 80
Toteşti-baraj, 54
Traianus, 5
Transylvania, 64, 84; climate of, 99, 101, 102; fauna of, 141, 143, 150; floras of, 102–4; geologic history of, 94–97; and historical contingency, 167, 194; insularity of, 106–12, 118, 143–46, 167–68; paleogeography of, 98–105; paleosols of, 101, 102
Transylvania Depression, Basin, 15, 25, 46, *94*, 98, 101, 105
tree ferns, 103
Tretosternon, 180
Triceratops, 23
Troodon, 30
Troodontidae, 31, 33, 36, 37
turtle, 73–77, 107–9, 118, 141, 184
turtle *(Kallokibotion)*, 11, 180
turtle, Galapagos, 119
Tuştea, 54, 56, *57*
typostrophy, 142
Tyrannosauroidea, 31, 110
Tyrannosaurus, 19, 30, 110, 117
Tyrannosaurus rex, 23, 124

Ukraine, 108
United States, 27, 117, 177
Urodela, 80

Valdosaurus, 169
Vălioara, 15, 60, 63, 67

Van Valen, Leigh, 143, 195, 196, 197
Vardar Ocean, 95
Velociraptor, 30, *32*, 36
Versluys, Jan, 10
Vertebrata, 19
Villaggio del Pescatore, 172
Vinţu de Jos, 15, 46
Vurpăr, 15, 40, 46, 73, 77, 101
Vurpăr Formation, 47

walnut (Juglandaceae), *103*, 104
Wegner, Alfred L., 89–91
whiptail lizards, 73
William of Ockham, 21
Wiman, Carl, 10
woodchuck, 77
worm lizards, 73

Zalmoxes, 2, 3, 43–48, 50, 59, 168, 176; body size of, 187; ghost lineage of, 157–59, 161; growth and development of, 126, 137–40, 146; migration of, 182, 183; and Nopcsa, 10; and North America, 177; and rate of evolution, 160–62; source area for, *178*
Zalmoxes robustus, 16, 46, 138, 157, 159, 160; growth and development of, 125, 139–41, 184
Zalmoxes shqiperorum, 46, 157, 159, 160; growth and development of, 138–40, 184
Zeicani, 87
ziphodonts, 67
Ziphosuchia, 71